This book provides a brief but thorough account of the basic principles of good pump design.

The first three chapters of the book present the basic hydraulic equations, including cavitation, and discuss the principles that underlie the correct performance of the impeller and casing of centrifugal pumps and the blading of axial machines. The fourth chapter outlines analytical methods for flow calculations, including special techniques used in computer aided design. The following two chapters work through two examples: the first the preliminary design of the impeller and casing of a simple volute type centrifugal pump; the second outlines alternative approaches to designing an axial flow pump impeller. The important topics of shafts, bearings, seals and drives are discussed in a separate chapter, followed by a short chapter on the problems associated with designing pumps for use with difficult liquids. The final chapter introduces the industrial codes and practices that must also be taken into account in finalising any pump design.

This text will be of interest to graduate students, research and professional designers in mechanical, aeronautical, chemical and civil engineering.

T0282335

ROTODYNAMIC PUMP DESIGN

ROTODYNAMIC PUMP DESIGN

R. K. Turton
Loughborough University of Technology

CAMBRIDGE UNIVERSITY PRESS
Cambridge, New York, Melbourne, Madrid, Cape Town, Singapore, São Paulo

Cambridge University Press
The Edinburgh Building, Cambridge CB2 2RU, UK

Published in the United States of America by Cambridge University Press, New York

www.cambridge.org
Information on this title: www.cambridge.org/9780521305020

First published 1994
This digitally printed first paperback version 2005

A catalogue record for this publication is available from the British Library

Library of Congress Cataloguing in Publication data
Turton, R. K. (Robert Keith)
Rotodynamic pump design / R. K. Turton.
     p.   cm.
Includes bibliographical references and index.
ISBN 0–521–30502–0
1. Rotary pumps – Design and construction.   I. Title.
TJ917.T87   1993
621.6´6 – dc20   93–13772   CIP

ISBN-13 978-0-521-30502-0 hardback
ISBN-10 0-521-30502-0 hardback

ISBN-13 978-0-521-01962-0 paperback
ISBN-10 0-521-01962-1 paperback

Contents

Preface   *page* xi
Symbols   xiii

1 Fundamental principles   1

    1.1   Introduction   1
    1.2   A black-box approach   1
    1.3   The Euler equation   2
    1.4   The ideal equations for a radial pump   3
    1.5   The ideal equations for the axial flow pump   4
    1.6   A dimensionless approach   5
    1.7   Specific speed   9
    1.8   Losses and loss estimation   10

2 Cavitation in pumps   15

    2.1   Introduction   15
    2.2   Bubble inception and collapse   15
    2.3   The effects of cavitation on pump behaviour   17
    2.4   Cavitation criteria   19
    2.5   Design criteria   24
    2.6   Scaling of cavitation   26
    2.7   The inducer   27

3 Centrifugal pump principles   29

    3.1   Introduction   29
    3.2   Choice of rotational speed   30
    3.3   Inlet design   31

3.4 The impeller 35
3.5 The collector system 47
3.6 Thrust loads due to hydraulic effects 53

4 Principles of axial and mixed flow pumps 60
4.1 Introduction 60
4.2 The isolated aerofoil concept 62
4.3 Blade data for axial flow machines 65
4.4 An approach to mixed flow machines 76
4.5 Thrust loads 79

5 Flow calculations in pumps and an introduction to computer aided techniques 80
5.1 Introduction 80
5.2 Stream-surfaces 80
5.3 Empirical techniques 82
5.4 Computer based theoretical techniques 82

6 Single stage centrifugal pump design 95
6.1 Introduction 95
6.2 Initial calculations 95
6.3 Suction geometry 98
6.4 First calculation of flow path 99
6.5 Approaches to outlet geometry determination 100
6.6 Calculation of outlet diameter and width using the 'slip' method 107
6.7 A discussion of blade design considerations 108
6.8 The outline impeller design 114
6.9 The design of the volute 117
6.10 Radial and axial thrust calculations 119
6.11 Wear ring design 120
6.12 Multistage pump design 122
6.13 Conclusion 122

7 The design of axial and mixed flow pumps 124
7.1 Introduction 124
7.2 An approach to design 125
7.3 Axial flow pump design, an empirical approach 127
7.4 Mixed flow pumps 138

8 Basic design principles of shafts, bearings and seals, and selection of drive 140

8.1   Introduction   140
8.2   Shaft design   140
8.3   Bearings   142
8.4   Seals   147
8.5   Selection of pump drives   157

9 Pump design for difficult applications   161

9.1   Introduction   161
9.2   Conventional process pumps   161
9.3   Solids handling pumps   168
9.4   Glandless pumps   168
9.5   Gas–liquid pumping   171
9.6   Conclusion   179

10 An introduction to the next stage in the pump design
process   180

10.1 Introduction   180
10.2 Establishing the pump duty   180
10.3 The design process   181
10.4 The influence of dimensional standards   181
10.5 The implications of API 610   182
10.6 Concluding remarks   185

References   187
Index   195

Preface

This text is intended as an introduction to rotodynamic pump design. Any successful pump must satisfy the following objectives:

It must give specific pressure rise and flow rate within acceptable limits at an acceptable rotational speed, and take minimum power from its drive; it must give a stable characteristic over the operating range required and while meeting all performance criteria, the cavitation behaviour must be good. The pump must be as small as possible, the power absorbed must normally be non-overloading over the flow range and the noise and vibration must be within specified limits. The design must always be economical, give good quality assurance, and be easily maintained.

In addition to the objectives stated, special requirements also have an influence on design. For example, pumps handling solids must resist erosion and blockage of flow passages. In many fluid processes the pumps have to cope with the multiphase fluids and high gas content. Modern boiler feed pumps pose particular problems of shaft and drive design. This text introduces the reader to design approaches which can deal with these and other problems.

It has been assumed that the reader has a basic understanding of fluid mechanics, so the treatment commences with a statement of the fundamental Euler equation and its applications, and continues with a fairly comprehensive discussion of cavitation, its effects, and basic design data relevant to rotodynamic pumps. The text then describes the fundamental design principles and information available on centrifugal and axial/mixed flow machines. A brief discussion of computer aided techniques is included. This is not a comprehensive text, but is instead intended to focus on fundamentals and to thus serve as an introductory treatment.

Hydraulic design approaches to radial and axial machines are outlined in Chapters 6 and 7, using worked examples to allow the reader to follow the design process.

Since the liquid end cannot operate without a drive provided with the necessary seal and bearing systems, Chapter 8 provides a summary of the basic principles followed in providing a total pump fit for the duty demanded.

The early chapters all assume a clean single phase liquid, so Chapter 9 discusses in some detail the problems posed by process chemicals which are aggressive, by solids in suspension, and the presence of gas in the pumpage. Solutions are described and, where space does not allow for full discussion, adequate references are provided for further reading.

All the experience of this designer and that of others, suggests that even in this computer age the design of a good pump is a combination of art and technology, and it is hoped that this distillation of the experience of many engineers proves to be of help for those wishing to design pumps, or to users who wish to understand how a pump design evolves.

No contribution to pump design can claim to be completely original, and the Author freely acknowledges the influence of many colleagues. All sources are acknowledged where possible, and the Author will be grateful for corrections and suggestions.

I gratefully acknowledge the skill of Janet Redman who translated my sketches, and I could not have completed my writing as effectively without the skills of Helen Versteeg and Gail Kirton who typed and laid out the original manuscript and improved my spelling. My patient and long-suffering wife June has helped with proofreading and coffee at appropriate intervals, and I dedicate this work to her.

Symbols used: their definitions and dimensions

a	acoustic velocity	$\mathrm{m\,s^{-1}}$
b	passage height	m
C_L	lift coefficient	
C_D	drag coefficient	
D	diameter	m
D	drag force on an aerofoil	N
F_a	force acting in the axial direction on a foil or blade	N
F_t	force acting in the tangential direction on a foil or blade	N
g	acceleration due to gravity	$\mathrm{m\,s^{-2}}$
gH	specific energy	$\mathrm{J\,kg^{-1}}$
h	specific enthalpy	$\mathrm{J\,kg^{-1}}$
H	head	m of liquid
K	lattice coefficient	
k_s	dimensionless specific speed	
L	lift force on an aerofoil	N
M	pitching moment acting on a foil	N m
M_n	Mach number $(=V/a)$	
$\dot{m}$	mass flow rate	$\mathrm{kg\,s^{-1}}$
N	rotational speed	$\mathrm{rev\,min^{-1}}$
NPSE	net positive suction energy	$\mathrm{kJ\,kg^{-1}}$
$\mathrm{NPSE_a}$	net positive suction energy available	$\mathrm{kJ\,kg^{-1}}$
$\mathrm{NPSE_R}$	net positive suction energy required	$\mathrm{kJ\,kg^{-1}}$
NPSH	net positive suction head	m of liquid
N_s	specific speed	

p	pressure	$N\,m^{-2}$
p_0	stagnation pressure	$N\,m^{-2}$
p_v	vapour pressure	$N\,m^{-2}$
P	power	$J\,s^{-1} = W$
Q	volumetric flow rate	$m^3\,s^{-1}$
R_e	Reynolds number	
R_{em}	model Reynolds number	
S	suction specific speed	
t	blade thickness	m
t	blade passage minimum width or throat	m
T	temperature (absolute)	K
T_0	stagnation temperature (absolute)	K
T	torque	$N\,m$
u	peripheral velocity	$m\,s^{-1}$
V	absolute velocity	$m\,s^{-1}$
V_a	axial component of absolute velocity	$m\,s^{-1}$
V_n	normal component of absolute velocity	$m\,s^{-1}$
V_R	radial component of absolute velocity	$m\,s^{-1}$
V_u	peripheral component of absolute velocity	$m\,s^{-1}$
W	relative velocity	$m\,s^{-1}$
W_u	peripheral component of relative velocity	$m\,s^{-1}$
Z	height above datum	m
α	angle made by absolute velocity	degrees
β	angle made by relative velocity	degrees
γ	ratio of specific heats	
γ	stagger angle	degrees
δ	deviation angle	degrees
ε	fluid deflection	degrees
ζ	loss coefficient (Equation 3.13)	
η	efficiency	
θ	camber angle	degrees
κ	elastic modulus	$kg\,m^{-1}\,s^{-2}$
μ	absolute viscosity	$kg\,m^{-1}\,s^{-1}$
ν	kinematic viscosity	$m^2\,s^{-1}$
ρ	density	$kg\,m^{-3}$
σ	Thoma's cavitation parameter	
ϕ	flow coefficient (V_a/u)	
ψ	specific energy coefficient	
Ω	Howell's work done factor	
ω	angular velocity	$rad\,s^{-1}$

Subscripts 1, 2 etc. indicate the point of reference.

1 □ Fundamental principles

1.1 Introduction

The same fundamental principles apply to most turbomachines and are well covered in texts dealing with turbomachines, like Wislicenus (1965) Dixon (1975) and Turton, (1984) but it is necessary to restate them here as an introduction to the design parameters which depend upon them. In this chapter, the basic equations relating the specific energy rise to flow rate to the basic geometry of the machine and to its physical size are stated, the normal non-dimensional and dimensional parameters are introduced and the discussion ends with a brief treatment of losses that occur in pumps with an introduction of some of the correction equations used.

1.2 A black-box approach

Consider a simple pump, Figure 1.1, through which liquid is passing and into which power is being transferred. By considering the pump as a control volume or a 'black box' we can examine the fluid energy change between the inlet and outlet planes and relate this to the energy input. Applying the energy equation it can be stated that,

$$\omega T = \left(\frac{P_2 - P_1}{\rho}\right) + \left(\frac{V_2^2 - V_1^2}{2}\right) + g(Z_2 - Z_1) + \text{losses} \tag{1.1}$$

The overall pressure rise experienced by the fluid is clearly a function of the internals of the box and their operation on the fluid must now

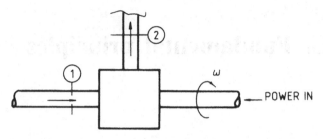

Figure 1.1. A black-box representation of a pump.

be discussed in some detail. In this introduction the ideal situation will be discussed and later chapters will give the detail necessary to allow for the real effects of three-dimensional flow and other fluid dynamic problems.

1.3 The Euler equation

This equation deals with an ideal fluid passing through an impeller, such as that shown in Figure 1.2. The impeller produces a form of forced vortex, and there is an exchange of angular momentum between the blades of the rotor and the liquid. Since it is well covered in the texts already cited, the two working equations will be stated

$$T = \dot{m} \left[Vu_1 \frac{D_1}{2} - Vu_2 \frac{D_2}{2} \right] \tag{1.2}$$

from which since the power input is ωT it can be shown that specific energy change per unit mass,

$$gH_E = u_1 Vu_1 - u_2 Vu_2 \tag{1.3}$$

This is known as the Euler equation, and the velocity components Vu_1 and Vu_2 are usually called whirl velocities. By using Pythagoras, Equation (1.3) can be rewritten to read

$$gH_E = \tfrac{1}{2}[\underset{(1)}{(V_2^2 - V_1^2)} + \underset{(2)}{(u_2^2 - u_1^2)} + \underset{(3)}{(W_1^2 - W_2^2)}] \tag{1.4}$$

Where term 1 is the change in absolute kinetic energy through the impeller, term 2 is a statement of the effect of radius change, and term 3 is the change of relative kinetic energy through the rotor. The latter two terms constitute the static energy change through the impeller,

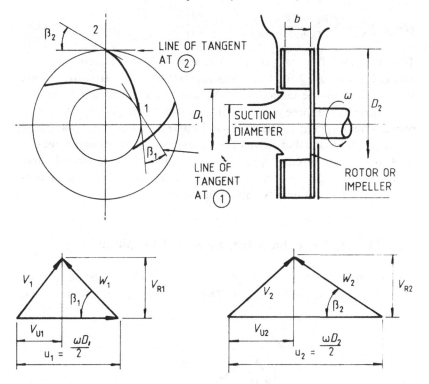

Figure 1.2. A simple centrifugal pump impeller, with the ideal velocity diagrams.

since gH_E is total energy change. This statement allows us to examine the relative role of diameter and related geometric changes on the energy increase produced by an impeller.

1.4   The ideal equations for a radial pump

If the impeller in Figure 1.2 and the resulting velocity triangles also in Figure 1.2 are examined again, but with the zero inlet whirl condition giving a right angled inlet triangle, the so-called ideal design condition, the Euler Equation will read $gH_E = u_2 Vu_2$ and by substitution

$$gH_E = u_2(u_2 - V_{R_2} \cot \beta_1) \qquad (1.5)$$

this can be transformed (noting that $V_{R_2} = Q/\pi D_2 b$) into the equation

$$gH_{\mathrm{E}} = u_2^2 - \frac{Q \cdot u_2 \cdot \cot \beta_2}{D_2 b} \tag{1.6}$$

Thus when the flow rate is zero (the shut valve condition), $gH_{\mathrm{E}} = u_2^2$ and gH_{E} varies as Q for any given geometry. It is instructive to look at the effect of the choice of outlet angle, and Figures 1.3 and 1.4 illustrate this, the three possible configurations being shown. The most conventional system is the backward curved layout, giving a falling characteristic from shut valve. The other systems are possible but not used usually for pumps. Also shown in Figure 1.4 is a typical curve illustrating how the usual curve differs from the ideal. These differences are discussed in Section 1.6 and Chapter 3.

1.5   The ideal equations for the axial flow pump

An axial flow pump of the simplest design with no inlet or outlet rows of blades is shown in Figure 1.5. The ideal flow situation here is that

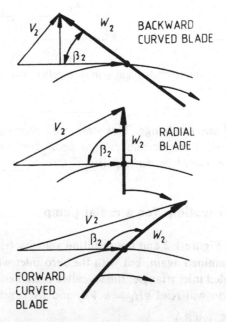

Figure 1.3.   The effect of Blade outlet angle β_2 on the outlet velocity diagrams for a centrifugal pump.

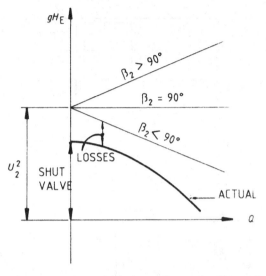

Figure 1.4.   The effect of blade outlet angle β_2 on the ideal specific energy rise produced by a centrifugal impeller.

flow proceeds along axysymmetric flow surfaces, and that the velocity at inlet is axial and uniform across the annulus. The Euler equation reduces, since the peripheral velocity remains constant through the blade row, to

$$gH_E = u[Vu_2 - Vu_1] = u\Delta Vu \qquad (1.7)$$

or in an extended form of Equation (1.4)

$$gH_E = \tfrac{1}{2}[(V_2^2 - V_1^2) + (W_1^2 - W_2^2)] \qquad (1.8)$$

The velocity triangles for the ideal flow situation are also shown in Figure 1.5. both as separate diagrams and on a common base which highlights the change in ΔVu. In axial pumps the change in specific energy is small so that the ΔVu component tends to be small, and the blade angles are low referred to the tangential direction.

1.6   A dimensionless approach

Reverting to the 'black-box' concept used earlier, it can be argued that the fluid power input is related to the flow rate Q, the specific energy

Fundamental principles

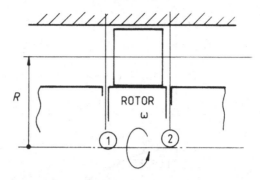

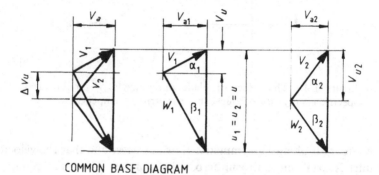

COMMON BASE DIAGRAM

Figure 1.5. A simple axial flow pump with the associated velocity
diagrams.

change gH_e, the fluid properties μ and ρ, the physical size of the
machine characterised by a diameter D, and to the rotation speed ω.
Applying the principle of dimensional analysis it can be shown that the
following relation will emerge:

$$\frac{P}{\rho\omega^3 D^5} = f\left[\frac{Q}{\omega D^3} \cdot \frac{gH}{\omega^2 D^2} \cdot \frac{\rho\omega D^2}{\mu} \ldots\right] \tag{1.9}$$

The LH group is a power coefficient, the first group on the RH side
can be written as a specific energy coefficient by remembering that
ωD can be written as u so that $\psi = g(H/u^2)$ (the non SI statement is
known as the head coefficient). Similarly the second term of the RH
side can be transformed into a flow coefficient $\phi = V_a/u$, and the
third is a form of Reynolds number, which may be written as $\rho u D/\mu$
or $\rho\omega D^2/\mu$.

These groups are non-dimensional, and can be used to plot machine performance data for a family of machines that are all similar, and also to predict the performance of a similar machine if that of one size is known, and as a basis for storing design information. Figure 1.6 illustrates the way a set of dimensional curves can be presented as one curve set.

Performance prediction or scaling laws can be used for groups of similar machines following the rules based on Equation (1.9).

$$
\left.
\begin{aligned}
\frac{Q}{\omega D^3} &= \text{constant} \\[2ex]
\frac{gH}{\omega^2 D^2} &= \text{constant} \\[2ex]
\frac{P}{\rho \omega^3 D^5} &= \text{constant}
\end{aligned}
\right\}
\qquad (1.10)
$$

Figure 1.7 indicates how the effects of speed change can be predicted using Equations (1.10). For example take the performance points for flow Q_A: using Equations (1.10), flow Q_B at the lower speed ω_B, gH_B, and power P_B may be found and the corresponding performance points found as shown in the figure.

An approach to the prediction of performance known as the 'scaling laws' is often used to correct for the effect of turning down or reducing the diameter of an impeller to reduce performance. The laws are usually written:

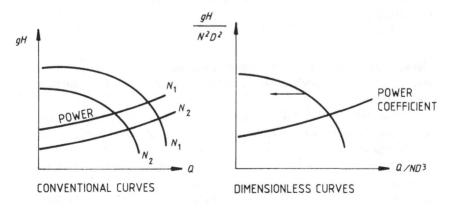

Figure 1.6. A non-dimensional performance plot.

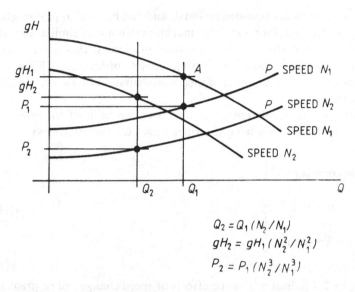

$$Q_2 = Q_1 (N_2 / N_1)$$
$$gH_2 = gH_1 (N_2^2 / N_1^2)$$
$$P_2 = P_1 (N_2^3 / N_1^3)$$

Figure 1.7. The prediction of the effect of rotational speed change using equation (1.10).

$$\left. \begin{array}{l} \dfrac{D_2}{D_1} = \dfrac{H_2}{H_1} \\[3mm] \dfrac{Q_2}{Q_1} = \sqrt{\left(\dfrac{H_2}{H_1}\right)} \\[3mm] \dfrac{P_2}{P_1} = \sqrt[3]{\left(\dfrac{H_2}{H_1}\right)^2} \end{array} \right\} \qquad (1.11)$$

Figure 1.8 is an example of the use of these equations based upon the need to reduce the specific energy output from a pump, by reducing the diameter.

As Karassik *et al.* (1976) and Stepannof (1976) among others comment, the flow is not strictly similar in the full size impeller and the turned down unit, so that there are differences, with the best efficiency point moving to lower flow rate than predicted with a small error, shown in Figure 1.8. The correction is performed using approaches like those outlined in Stepannof (1976), or similar adjustments based upon experience.

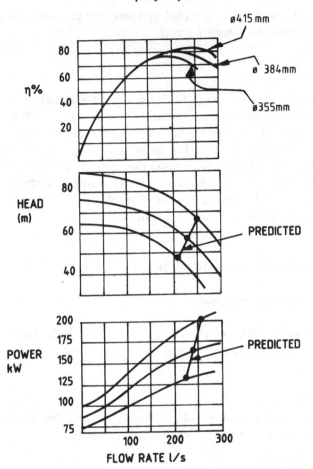

Figure 1.8. The use of the scaling laws to predict the change in performance resulting from diameter reduction.

1.7 Specific speed

A much used and well known parameter that was introduced many years ago, a statement based upon the head and flow at best efficiency point at design speed is the specific speed.

In its non-SI form the specific speed is defined as

$$N_s = \frac{N\sqrt{Q}}{H^{3/4}} \tag{1.12}$$

In the SI form the term is called variously the characteristic number or non-dimensional speed number

$$k_s = \frac{\omega \sqrt{Q}}{(gH)^{3/4}} \tag{1.13}$$

The form used in Equation (1.12) is not non-dimensional, and is quoted in a number of dimensional systems, from the rpm – US gallon per minute – foot used in the USA, to the cubic metre per second – rpm – metre head used in some European references, so care has to be taken when data is used based upon this number. Figure 1.9 indicates how flow path, general cross-section configuration and characteristics relate to k_s, and Table 1.1 gives multiplying factors that convert conventional specific speeds to k_s values.

1.8 Losses and loss estimation

1.8.1 Efficiency statements

If the actual specific energy rise is compared with the Euler specific energy change the ratio is known as the Hydraulic Efficiency

$$\eta_H = \frac{gH}{gH_E} \tag{1.14}$$

If the mechanical and non-flow losses are included the resulting efficiency is the overall efficiency

$$\eta_o = \eta_H \times \eta_{MECH} \tag{1.15}$$

A much quoted plot of η_o against N_s is found in many texts and Figure 1.10 illustrates this plotted against the SI number. It allows both the efficiency to be estimated and the relation of the flow path shape to this number to be examined. Low numbers indicate a high specific energy change and a low flow (associated with a purely radial flow impeller). A high number indicates a high flow rate and a low specific energy change, and consequently an axial flow design.

1.8.2 Loss estimation

Losses that contribute to the efficiency of the pump may be categorised in the form of an equation;

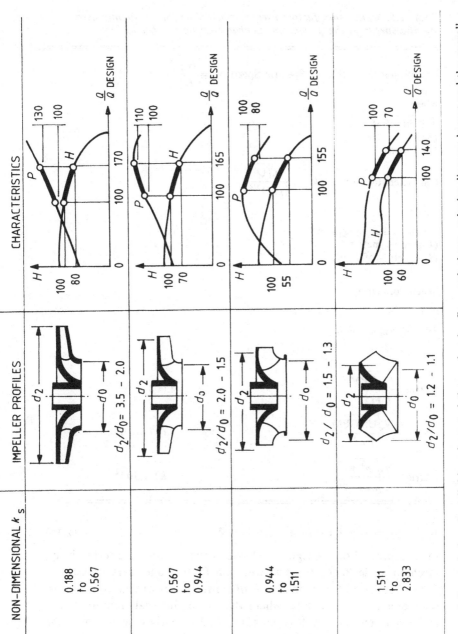

Figure 1.9. A design chart linking characteristic number k_s, flow path shape, velocity diagram layout and the overall characteristics.

Table 1.1. *Conversion factors used to convert several commonly used*
specific speeds to the dimensionless characteristic number k_s.

Traditional Definition of Specific Speed $N_S = \dfrac{N\sqrt{Q}}{H^{3/4}}$

where

N is in RPM
Q is flow rate
H is head

Characteristic Number $k_S = \dfrac{\omega\sqrt{Q}}{(gH)^{3/4}}$

where

ω is in radians per second
Q is cubic metres
H is metres of liquid
g is $9.81\ \mathrm{m\,s^{-2}}$

When converting:

$N_S K = k_S$

Typical Values of K:

	N_S	K
British	$\dfrac{N\sqrt{\mathrm{GPM}}}{\mathrm{Ft}^{\frac{3}{4}}}$	4.008×10^{-4}
US	$\dfrac{N\sqrt{\mathrm{US\,GPM}}}{\mathrm{FT}^{\frac{3}{4}}}$	3.657×10^{-4}
Metric	$\dfrac{N\sqrt{m^3/s}}{m^{\frac{3}{4}}}$	1.89×10^{-2}

$$(1 - \eta_T) = \delta_T = \delta_M + \delta_L + \delta_D + \delta_F + \delta_I \qquad (1.16)$$

those expressed on the right hand side of the equation are respectively, mechanical, leakage, disc friction, skin friction and inertia losses.

The mechanical loss, made up of power losses to bearings, and to seals and also to windage where this is not internal, represents the irreducible part of the loss equation, and is fairly constant over the usual range of speeds. The leakage loss is usually allowed for by using a volumetric efficiency and is due to the flow through the wear or

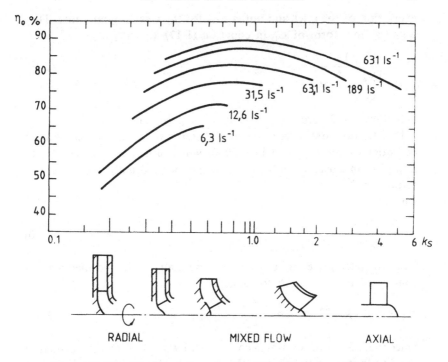

Figure 1.10. The variation in overall efficiency with the characteristic number k_s and with flow rate (after Turton, 1984).

check ring clearances of liquid from the impeller delivery pressurised region back to the suction. This flow has to be added to the delivered flow to give the flow rate that must be operated on by the impeller. One approach to estimation of the flow rate is to assume that the wear ring clearance acts as an annular orifice. Several formulae are to be found in the literature, Stepannof (1976) for example quotes a formula based on friction coefficients in Chapter 10 of his text. Denny (1954) did some tests and proposed

$$Q_L = C_D A_L \sqrt{2gH_L} \tag{1.17}$$

where

$$C_D = \frac{1}{[\lambda + (K_i + K_o)]} \tag{1.18}$$

Nixon and Cairney (1972) state an equation for λ based on several published sets of data, and comment that the term $(K_i + K_o)$ is an expression of the inlet and outlet loss for the wear ring annulus. They quote a final form of C_D in equation (1.17) as being

$$C_D = \frac{1}{\left(\dfrac{\lambda L}{2C} + 0.85\right)} \tag{1.19}$$

and show good agreement over a wide speed range.

Disc friction loss correlations in the literature are mostly based upon the work of Nece and Dailly (1960) who studied the torque due to thin discs in closed chambers, the torque being usually expressed as a torque coefficient C_m, where

$$C_m = \frac{2T}{\rho \omega^2 R^5} \tag{1.20}$$

Since the back plate and the shroud are usually machined they may be considered to be smooth for which

$$C_m = 0.075\, Re_D^{-0.2} \left[1 + 0.75\left(\frac{S}{R} + \frac{t}{R}\right)\right] \tag{1.21}$$

Since these relations do not account for flow from discharge to suction the disc friction is larger than they predict. More recent work, by for example Kurokawa (1976) indicates the influence of the through flow, and a working relation is to assume that the actual loss is of the order of 10% more than that predicted by Equations (1.20) and (1.21).

The flow losses may be considered to be due to skin friction and inertia effects. As work on water turbines as well as pumps has shown, the inertia losses tend to remain fairly constant over the flow range and they tend to be insensitive to Reynolds number effects. Skin friction follows Reynolds number, and the work published by Nixon and Cairney (1972) gives some very useful information on the effects of surface roughness and Reynolds number. Parallel work on fans published by Myles (1969), using a diffusion factor suggests an alternative approach. Clearly the estimation of pump loss is a complicated matter, and many engineers prefer the approach of Anderson for example where the experience over a long time is consolidated into a single chart and calculation. The contributions of Anderson (1977, 1980) are very good examples of this technique and should be considered. The many pump handbooks may also be consulted.

2 □ Cavitation in pumps

2.1 Introduction

Bubbles form in a flowing liquid in areas where the local pressure is close to the vapour pressure level. They form and collapse in a short time, measured in microseconds, and their life history gives rise to local transiently high pressures with flow instability. In pumps this results in noise, vibration and surface damage which can give rise to very considerable material loss.

The inception and collapse mechanisms are discussed briefly in this chapter, as are the conventional empirical rules used to ensure satisfactory pump behaviour. The chapter concludes with a discussion of the design rules to be followed in producing a good pump, and with a treatment of the techniques used to predict cavitation performance.

2.2 Bubble inception and collapse

In theory, cavities will form when the local liquid pressure level is equal to the vapour pressure under the local conditions. In practice bubbles form at higher pressure levels, due in part to the presence of very small bubbles or particles of detritus which act as triggers. A very exhaustive treatment of the process will be found in the monograph by Knapp *et al.* (1970), so a very brief summary will be given here.

Figure 2.1 is based on work done by Worster (1956) who used theoretical equations first published by Rayleigh (1917) to predict the life cycle of an existing small bubble as it grew and then collapsed.

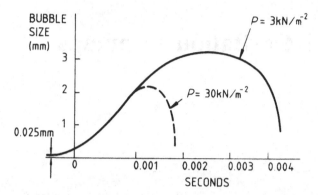

Figure 2.1.   The life history of a bubble for two collapse pressures
(based on work by Worster (1956)).

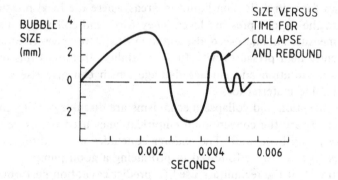

Figure 2.2.   The tendency for bubbles to collapse and reform
(based on work by Knapp and Hollander (1970)).

The figure illustrates the short life cycle, and Figure 2.2 demonstrates
the collapse and reform of bubbles that has been observed when the
mechanics of cavity behaviour have been studied. A full treatment
is not possible, so the reader is referred to the monograph by Knapp
et al. (1970), Lush (1987a and b), Durrer (1986) and others for a more
comprehensive discussion. As both Lush (1987b) and Durrer (1986)
among others, discuss cavitation erosion modelling may be done using
the concept of the micro-jet (Figure 2.3), examining the behaviour
of a bubble close to a surface, the micro-jet moving at velocities up

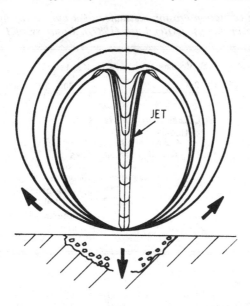

Figure 2.3.   The concept of the micro-jet formed when a bubble collapses (Lush (1987b), courtesy of the Institution of Mechanical Engineers.

to 400 ms^{-1}. This effect has been calculated to give rise to instantaneous pressures up to 6000 atmospheres on collapse, and local metal temperatures up to 10 000°C have been suggested. These values have been used to explain the typical erosion and corrosion in the area of cavitation bubble collapse. It is argued that initially the metal surface hardens and the locally high temperature affects the material just below the surface and makes it a little 'spongy', so that the surface cracks allow acid attack below the surface and thus the familiar lunar landscape results. Resistance to damage correlates with surface hardness, as Table 2.1 indicates.

2.3   The effects of cavitation on pump behaviour

Figure 2.4 demonstrates that the static pressure falls as the liquid nears the leading edges of the impeller blades, and continues to drop as flow takes place round the nose of the vanes. The pressure can drop

Table 2.1. *The cavitation erosion resistance of a number of materials commonly used in pumps. (After Lush (1987)). Courtesy of I. Mech. E.*

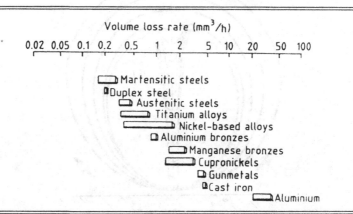

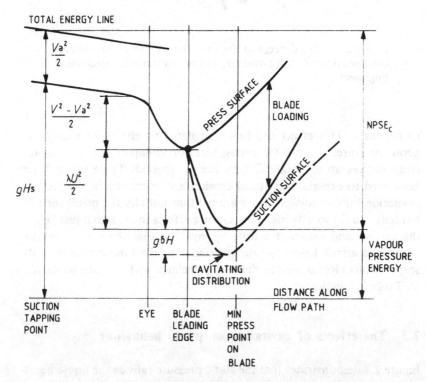

Figure 2.4. The variation of local pressure as flow approaches the impeller of a centrifugal pump.

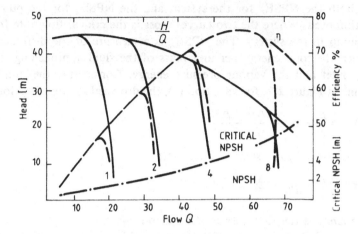

Figure 2.5. The effect of cavitation on a centrifugal pump characteristic.

to round vapour pressure level so cavities form and collapse in the passages of the impeller, causing flow instability as well as damage. The hydraulic effect is demonstrated in Figure 2.5, which shows how the flow range is limited progressively as the suction pressure presented to the impeller falls. This simple presentation is one way of presenting the effects, but the more common method is to use the concept of Net Positive Suction Energy NPSE (or as used for many years, Net Positive Suction Head NPSH) and pump testing codes usually ask for the variation of pump head with pump NPSH when running at constant flow and rotational speed.

Criteria for judging whether cavitation behaviour is acceptable or not will now be discussed.

2.4 Cavitation criteria

As the name suggests, NPSH was defined as the head above vapour pressure head found in the suction area of the pump. With the advent of the SI system this is now defined as the difference between the suction pressure energy and the vapour pressure energy, expressed in absolute terms. Two definitions are used, NPSE available ($NPSE_A$), and NPSE required ($NPSE_R$). Figure 2.6 shows a pump characteristic with both the NPSE_A for the system and the NPSE_R for the pump superimposed. Where the two curves cross is the critical flow rate for the pump in the system. The NPSE_A is a measure of the difference between the total energy per unit mass of the fluid approaching the pump inlet and the vapour pressure energy. Considering the simple systems in Figure 2.7, for variation (a), the drowned suction situation,

$$\text{NPSE}_A = \frac{p}{\rho} + gh_1 - \frac{\Delta p}{\rho} - \frac{p_v}{\rho} \qquad (2.1)$$

For variation (b) the suction lift situation,

$$\text{NPSE}_A = \frac{p}{\rho} - gh_1 - \frac{\Delta p}{\rho} - \frac{p_v}{\rho} \qquad (2.2)$$

where $\Delta p/\rho$ is the flow loss in the suction line.

Figure 2.8 illustrates the variations imposed by suction system conditions on the NPSE_A.

The pump by its own dynamic action can sustain a suction pressure and this value is covered by the term $\text{NPSE}_{\text{required}}$ (NPSE_R) used by the pump manufacturer. This is a function of the pump geometry, and is conventionally presented as Figure 2.9, where the pump energy rise is shown plotted against NPSE_R for a machine rotating at constant speed and passing a constant flow rate. The so-called critical value of NPSH_R is defined in the figure, the point on the curve at which the energy rise falls by $X\%$ from the design or duty level. A term

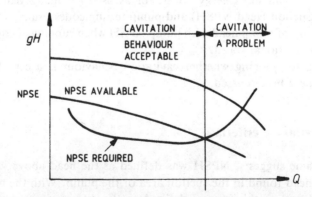

Figure 2.6. A pump characteristic with the system NPSE_A and Pump NPSE_R curves superimposed to illustrate their interrelation.

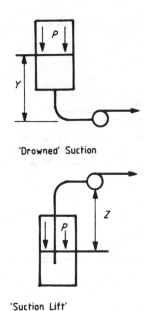

(a) 'Drowned' Suction

(b) 'Suction Lift'

Figure 2.7. The two main types of suction system encountered by
pumps.

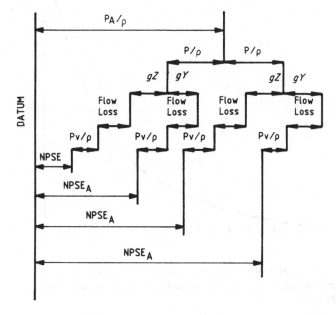

Figure 2.8. The NPSE available at the pump calculated for four
typical operating situations.

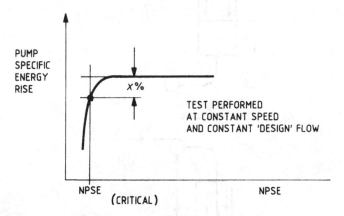

Figure 2.9.   The conventional way in which the critical NPSE for a pump is presented.

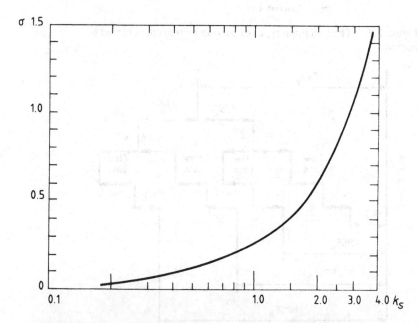

Figure 2.10.   A plot of the Thoma cavitation parameter against k_S (based on the curve in the American Hydraulic Institute Standards (1983)).

originally derived by water turbine manufacturers is much used: the
Thoma Cavitation parameter σ, defined as

$$\sigma = \frac{\text{Pump Energy Rise}}{\text{NPSE}_R} \qquad (2.3)$$

Another commonly used parameter has been suction specific speed,
k_{ss} being the SI form, defined as

$$k_{ss} = \frac{\omega\sqrt{Q}}{(\text{NPSE}_R)^{3/4}} \qquad (2.4)$$

Typical values for a pump with a mid-range characteristic number
are 3, (7500 (ft gpm rpm)), for a 'good' centrifugal pump 3.8 to 4.2
(9500–10 500) and for condensate and double suction centrifugal
pumps 5 (12 500).

A replot of the conventional presentation of σ against the non-
dimensional speed k_s is shown as Figure 2.10 which is based on the
American Hydraulic Institute Standards (1983). Prediction of the
critical point is difficult, and a recent approach based on noise level
and maximum material removal indicates one monitor level worth
considering. Figure 2.11, based on studies performed at the National

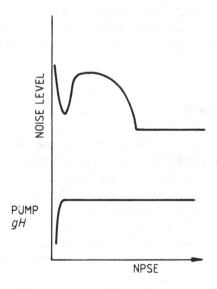

Figure 2.11. The variation of the noise generated by a pump
plotted against NPSE (from Deeprose (1977)).

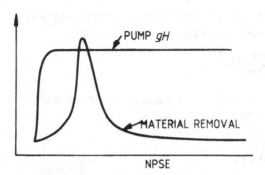

Figure 2.12. Metal removal from an aluminium specimen by cavitation presented against noise level (based on N.E.L. work presented by Pearsall (1974)).

Engineering Laboratory (Deeprose, 1977) indicates the relation between noise level and the NPSE curve. Figure 2.12 shows diagrammatically the relation between material loss and noise level, and indicates the validity of noise level monitoring as an indicator of material loss and cavitation problems.

While this sort of data is an indicator, and Figure 2.10 or similar charts that may be found for example in Wisclicenus (1965) may be used, the designer really needs more data to determine the correct inlet geometry, and approaches to this will be described in the following section.

2.5 Design criteria

Approaches to design are to be found in the contributions of Pearsall (1973), Anderson (1980), Gongwer (1941), Lewis (1964), and Ward and Sutton (1958) among others. Pearsall developed a design approach to optimising the inlet diameter using work by Gongwer among others and Figure 2.13 comes from his paper. He discusses the relation between axial velocity and cavitation, and the following derivation and discussion is a summary of a longer treatment.

Forming an expression for $NPSE_R$ from Pearsall, Lewis and Gongwer

$$NPSE_R = 1.8 \frac{V_{m1}^2}{2} + 0.23 \frac{U_{1e}^2}{2} \qquad (2.5)$$

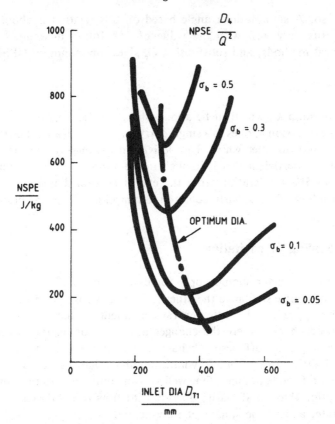

Figure 2.13. The design chart for optimum suction diameter for
good cavitation performance of a centrifugal pump (Pearsall,
1973), courtesy of the Institution of Mechanical Engineers.

where $V_{m1} = Q/$(suction area), and U_{1e} is peripheral velocity calculated
for the maximum inlet diameter. For 'critical' operating conditions
in the pump inlet, both the velocities used are based on the eye diameter
D_E, so differentiation and equating to zero gives an optimum value
for the eye diameter

$$D_E = K \left(\frac{Q}{\omega}\right)^{1/3} \tag{2.6}$$

where $K \simeq 4.66$

Anderson (1980) shows that there is an optimum ratio that is of
the order of 4, and it can be shown from the above equations that

this is so. A suitable inlet angle based on this relation is about 14°, which fits very well with the value of 15° found in a number of empirical methods, and substituting D_e gives the minimum $NPSE_R$ as

$$NPSE_R = \frac{\omega^{4/3}}{Q^{2/3}}$$

The optimum k_{ss} will then be approximately 3.25.

This expression applies to a single entry impeller with an unobstructed suction and no inlet whirl. The derivation assumes a vane inlet tip cavitation coefficient that is implicit in the approach due to Gongwer and also 5% acceleration from the eye or suction diameter into the vane leading edge, a fairly common assumption in current designs.

2.6 Scaling of cavitation

Thoma's parameter should apply for dynamically similar conditions. This is of course similar to the other scaling laws covered in Chapter 1.

There appears to be little general agreement on the effects of size and speed change where the changes are large, or on the effects of these changes on off design behaviour. Deeprose and Merry (1977) showed little effect from Reynolds number changes but a definite trend with Froude number. One well known company uses the working assumption that NPSE varies linearly with flow below design or duty flow rate, and as the square of flow above.

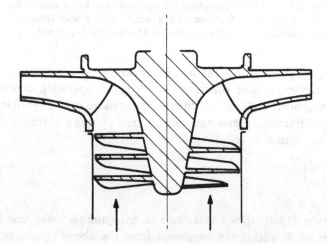

Figure 2.14. A centrifugal pump/inducer combination.

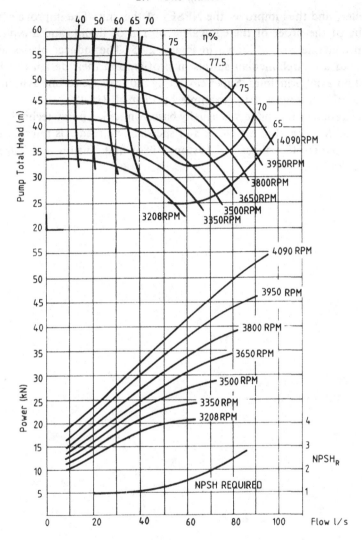

Figure 2.15. The performance of a centrifugal pump/inducer combination.

2.7 The inducer

An inducer is an axial type first stage to the impeller immediately up stream from the impeller of a centrifugal pump (Figure 2.14). This small pump increases the pressure in the suction area of the main impeller, and thus improves the $NPSE_R$ of the unit. The improvement can be of the order of the equivalent of 1 m of liquid, and a typical pump characteristic is shown in Figure 2.15. The inducer blades are few, and are Archimedian spirals in profile, and their design can be based on axial principles. Super cavitating inducers based on cavitating hydrofoils are outlined for example by Pearsall (1970).

The design of the inducer is usually based on rocket pump technology extended to normal pump flows and speeds. The reader is referred to general papers like those by Turton (1984) and Turton and Tugen (1984a and b).

3 □ Centrifugal pump principles

3.1 Introduction

When designing a pump a number of design variables need to be determined:

- impeller rotational speed
- impeller inlet or suction dimensions
- impeller outlet diameter
- impeller blade number
- impeller blade passage geometry, including inlet and outlet blade angles
- impeller position relative to the casing
- collector leading dimensions (volute throat area or diffuser geometry)
- pump construction and materials.

There are a number of approaches to design, chief among which are: small changes from existing designs to give a slight change in head or flow range; design using empirical information, tabular and graphical; and computer based approaches which are in some instances based on empirical data and more recently use finite element or finite difference approaches. The use of these techniques will be discussed later. The sections which now follow survey some of the empirical information available. Typical pump cross-sections of single-stage end suction, and double suction designs and of a multi-stage machine are shown in Figures 3.1, 3.2 and 3.3.

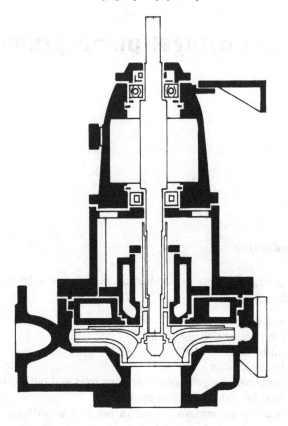

Figure 3.1.   A cross-section of a typical end suction 'back pull out'
centrifugal pump.

3.2 Choice of rotational speed

As will be clear from a reading of the later chapters on design, the
choice of rotational speed is interlocked with other parameters, but
there are empirical speed limits as given, for example, by the American
Hydraulic Institute Standards (1983) reproduced in many handbooks.
Clearly the rotational speed is limited to a range of synchronous speeds
when using electric motor on a 50 or 60 HZ supply frequency. For large
pumps, turbine or diesel drive is used, and the eventual rotational speed
is a compromise between hydraulic design and driver considerations.

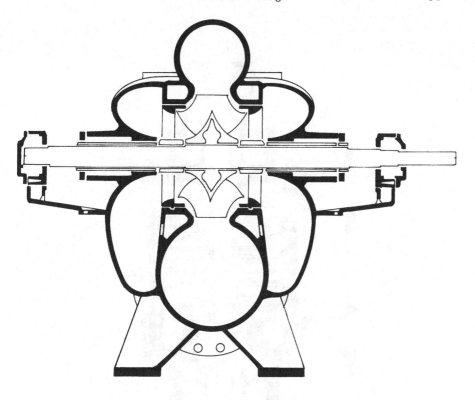

Figure 3.2.   A cross-section of a double suction centrifugal pump.

In general the value of the characteristic number k_s for rotodynamic pumps will be in the range 0.2 to 5, and if the initial driver speed chosen gives a value outside this range the choice should be reconsidered. Figure 3.4 shows how overall size reduces if the rotational speed can be increased without hydraulic penalty.

3.3   Inlet design

When a pump is being designed the optimum eye and inlet configuration must be determined. For an end suction design the optimum inlet is that provided by a straight inlet pipe (Figure 3.5(*a*)) as this offers the best inlet flow patterns. Often a larger inlet line than the pump

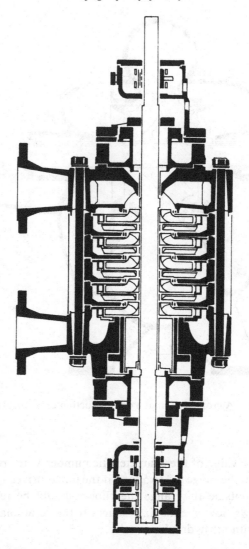

Figure 3.3.   A cross-section of a 5-stage barrel type pump.

eye diameter is provided either to reduce the suction line velocities
and hence losses or for installation reasons, and a conical reducer is
provided as shown in Figure 3.5(*b*). This will give a more confused flow
in the impeller eye, and in cases where there may be gas in the fluid,
or solids are being carried, an offset reducer as shown in Figure 3.5(*b*)

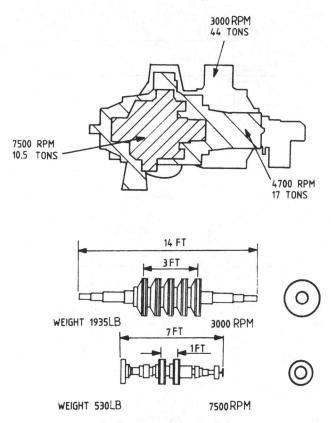

Figure 3.4. The effect of speed increase on the size and weight of a modern boiler feed pump.

is provided. This ensures that gas pockets do not form on shut down and also provides a route for solids to move away from the impeller when flow ceases. Figure 3.5(c) illustrates a bell mouth intake used when liquid is taken from a tank or vessel. The diameter D_0 is empirically determined from design charts as shown in Chapter 5 of Stepannof (1976), or by using the type of approach to limit cavitation as discussed in Section 2.5. Inlet zero whirl is assumed at design flow. Caution must be exercised in selecting a large inlet value of D_0 as, if it is too large, recirculation can occur. Figures 3.6 and 3.7 show the flow patterns that can arise at part flow as observed by Grist (1988). If D_0 is too large this recirculation can occur at a flow rate quite close

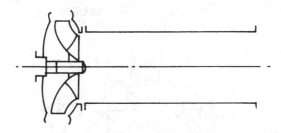

(a)

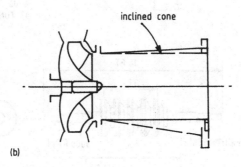

(b)

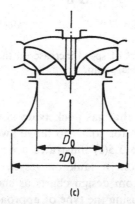

(c)

Figure 3.5.   Alternative suction systems for an end suction pump:
(a) A co-axial cylindrical straight suction line
(b) An inclined cone type suction
(c) A flared inlet that may be used for vertical designs.

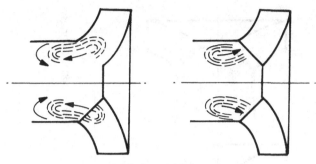

UNSTABLE PRE-ROTATION

Figure 3.6.   Unstable flow in the suction region of a centrifugal pump under 'part flow' operating conditions (after Grist (1988)), Courtesy of the Institution of Mechanical Engineers.

to the design value, and can result in an increase in the NPSH required as illustrated in Figure 3.8. If a mistake of this sort has occurred and tests indicate possible trouble due to recirculation an inlet 'bulk head' ring in the eye (Figure 3.9) can suppress the flow patterns at the expense of flow loss, so it is only a solution when the margin between $NPSH_R$ and $NPSH_A$ allows.

In many installations, for example most water supply pump stations, inlet bends must be provided. These give distorted flow patterns across the impeller eye. One design much used is shown in Figure 3.10(a). When the pump is of the double suction design typical suction duct layouts are as shown in Figures 3.10(b) and (c). Figure 3.11 gives an empirical design layout, and

$$D_1 = (1.02 \text{ to } 1.05) D_0 \tag{3.1}$$

$$d = (1.05 \text{ to } 1.10) D_h \tag{3.2}$$

and

$$D_s = (1.07 \text{ to } 1.11) (D_0^2 - d^2)^{\frac{1}{2}} \tag{3.3}$$

3.4   The impeller

3.4.1   General comments

The ideal relation between energy rise and impeller geometry is discussed in Section 1.4, and Equation (1.6) quotes the zero inlet whirl relation

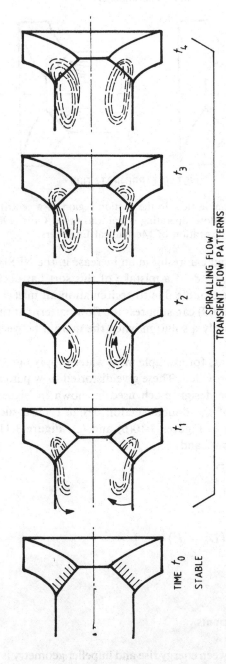

Figure 3.7. Transient flow patterns in the suction region of a pump at low flow rates (after Grist (1988)), Courtesy of the Institution of Mechanical Engineers.

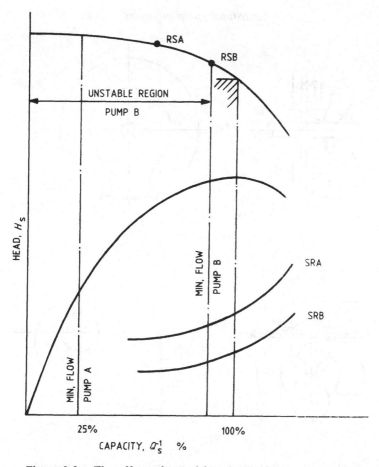

Figure 3.8. The effect of oversizing the suction on the NPSH (H_{SR}) and stability of two pumps. Pump *A* is the conventional suction size, and Pump *B* is the machine with an oversized suction.

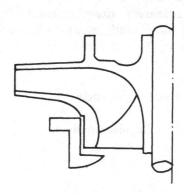

Figure 3.9. The use of an orifice plate to correct an oversized suction region.

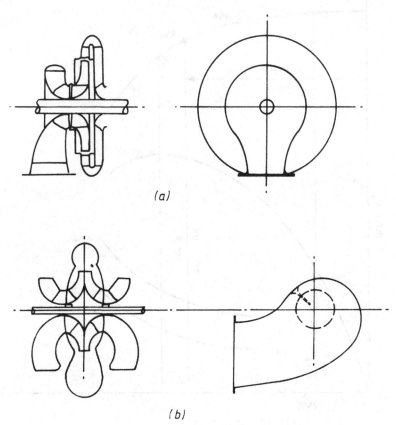

(a)

(b)

Figure 3.10.   Side channel designs of suction casing for (a) a single suction pump and (b) a double suction design.

for the Euler specific energy rise. From the Euler equation it may be deduced that the specific energy rise generated is a function of impeller outside diameter and outlet width, outlet angle and rotational speed at any given flow rate. Additional factors are blade or passage shape and blade number.

3.4.2   Blade number and outlet angle

Two simple design rules for blade number are

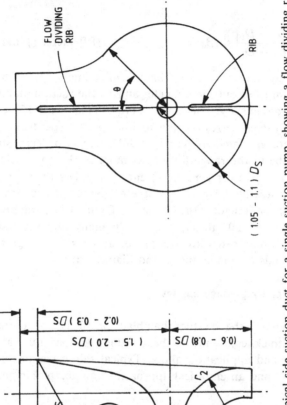

Figure 3.11. A typical side suction duct for a single suction pump, showing a flow dividing rib.

$$Z = \beta_2/3 \qquad\qquad \text{(Stepannof (1976))} \qquad (3.4)$$

or

$$Z = 6.5 \left(\frac{D_2 + D_1}{D_2 - D_1}\right) \sin \beta_m \qquad\qquad \text{(Pfleiderer (1961))} \qquad (3.5)$$

In these equations β_2 is the blade outlet angle, $\beta_m = (\beta_1 + \beta_2)/2$, D_2 is the impeller outside diameter and D_1 the suction diameter. Conventional designs use backward curved blades with $15^0 < \beta_2 < 35^\circ$, the lower values corresponding to low specific speed designs. These values were validated by Varley (1961) who also demonstrated that Z should be in the range 5 to 7, as more blades gave characteristics with a shut valve (or zero flow) energy rise less than the maximum value, and less blades gave unstable behaviour at low flow rates.

A further parameter is the overlap ϕ (Figure 3.12), and older designs used values of 180° in some cases. In many modern designs good efficiency is obtained with overlaps of about 45°, though the precise value depends on blade angles and diameter ratio.

3.4.3 Impeller passage shapes

Impeller passage shape rather than blade profile is an important factor, as blade thickness tends to be constant and as thin as strength, castability and application allow. Typical side elevations are seen in Figure 3.13 and an empirical approach to an acceptable passage area

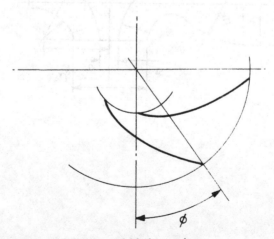

Figure 3.12.   A definition of blade overlap.

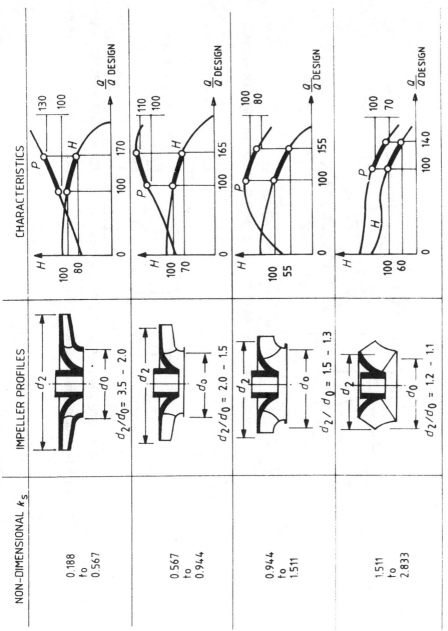

Figure 3.13. A presentation of the relation between impeller flow path shape and characteristics and the characteristic number k_s.

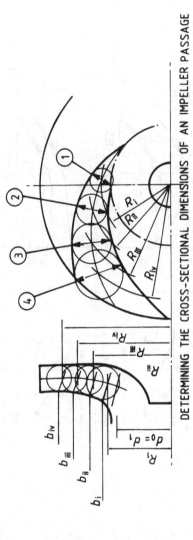

DETERMINING THE CROSS-SECTIONAL DIMENSIONS OF AN IMPELLER PASSAGE

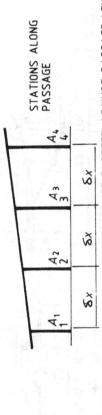

CROSS-SECTIONAL AREA OF IMPELLER PASSAGE PLOTTED AGAINST PASSAGE LENGTH

Figure 3.14. An empirical method of determining impeller blade passages suction of discharge.

change that ensures the outlet relative velocity is less than that at inlet is shown in Figure 3.14. The blade profiles generally follow an Archimedian spiral, with approximations using combinations of radii being used for drawing purposes as sketched in Figures 3.15 and 3.16. Where computer aided processes are used generating techniques have been developed, based on the point-by-point technique using a radius and reference angle as illustrated in Figure 3.17.

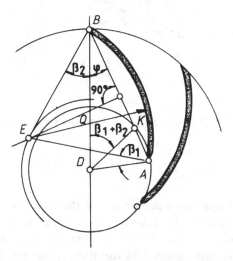

Figure 3.15. The single arc method of providing a blade profile.

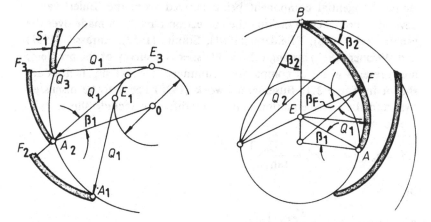

Figure 3.16. The double arc method of providing a blade profile.

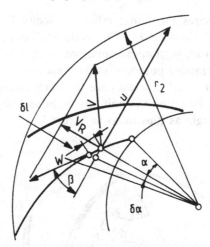

Figure 3.17. The point-by-point method of defining the blade profile.

3.4.4 The 'Slip' concept

Experimental studies have demonstrated that the flow both through the impeller passages and leaving the impeller is complicated and strongly three-dimensional. The outlet flow direction is consequently not the same as that assumed in the Euler Diagram. One approach designed to 'correct' for this flow distortion is the so-called 'Slip' or 'Head-Slip' approach. Figure 3.18 illustrates this technique with the 'actual' tangential component being derived from the 'Euler' value. Several approaches to making the correction have been made over the years: Peck (1968), Pfleiderer (1961), Stanitz (1952), Karassik (1981) and Weisner (1967) among others. Weisner is a survey of the problem, and presents a useful comparative summary. Several slip factors are shown in Table 3.1. Busemann's work (1928) needs to be discussed first and the Euler equation may be modified to include slip to take the form

$$gH = \eta_h u_2 \left(u_2 - W_e - \frac{V_{m_2}}{\tan \beta_2} \right) \tag{3.6}$$

or

$$gH = \eta_h u_2^2 \left[h_0 - \frac{V_{m_2}}{u_2 \tan \beta_2} \right] \tag{3.7}$$

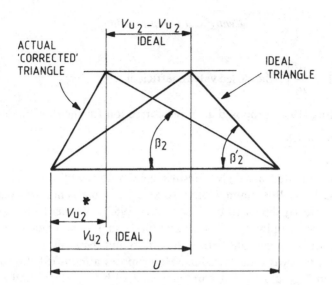

Figure 3.18. The concept of slip correction of the Euler triangle.

Table 3.1. *A summary of commonly used slip factors.*

Definitions of Slip Factor

Stanitz: $\mu = V_{U2/U_2} = 1 - 0.315 \left(\dfrac{2\pi}{Z} \sin \dfrac{\phi}{2} \right)$

ϕ is defined in Figure 3.20

Pfleiderer: $\dfrac{gH^*_{\text{Euler}}}{gH_{\text{Euler}}} = 1 + C_p$

gH^*_{Euler} corresponds to V^*_{u2} in Figure 3.18

$C_p = \left(\dfrac{(0.55 - 0.68) + 0.65 \sin \beta_2}{Z} \right) \dfrac{2 \times R_2^2}{(R_2^2 - R_1^2)}$

Weisner: $\sigma = \dfrac{V^*_{u2}}{u_2} + \dfrac{V_{R2}}{u_2} \tan \beta_2 = 1 - \dfrac{\pi \sin \beta_2}{Z}$

$$\text{or} = 1 - \dfrac{\sqrt{\sin \beta_2}}{Z^{0.7}}$$

This applies up to the limit $\varepsilon_{\text{Limit}}$ (see Figure 3.19)

of $R_1/R_2 = \left(\dfrac{1}{\log_e^{-1} \dfrac{8.16 \sin \beta}{Z} 2} \right)$

where

$$h_0 = \left(1 - \frac{W_e}{u_2}\right), \text{ the closed valve coefficient.}$$

Stodola (1945) proposed a simple correction for the flow distortion,

$$W_e = \frac{\pi \sin \beta_2 u_2}{Z} \tag{3.8}$$

This is still used as it gives results comparable with more complex formulations. Busemann (1928) conducted a mathematical analysis which is clearly discussed by Wisclicenus (1965). Weisner (1967) quotes the Wisclicenus plot of h_0 for several blade angles and blade numbers, and shows h_0 is constant with radius ratio and then falls as the radius increases towards outside radius. He proposes a modified slip factor, shown in Figure 3.19 that correlates well with Busemann, and shows his empirical formula is close to Busemann for conventional backward curved designs with the normal number of vanes. It may be commented that h_0 (or σ) is really a slip coefficient predicted for radial flow

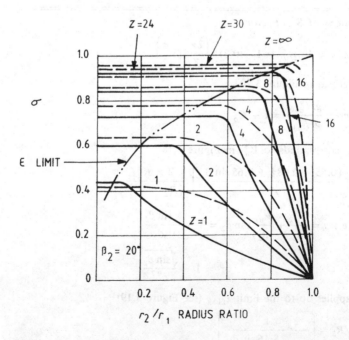

Figure 3.19.   The Weisner slip factor (based on Weisner (1967)).

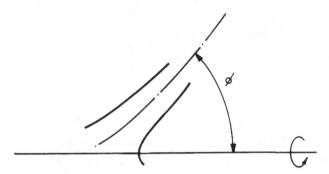

Figure 3.20. The 'layback' angle ϕ.

impellers, logarithmic spiral shaped blades, if the radius ratio is not small. If a mixed flow impeller is used a correction must be made for the 'layback' of the impeller ϕ defined in Figure 3.20. This layback is usually defined on the mid-stream surface, taken as an average for all stream surfaces, and h_0 for a radial machine is corrected using

$$h_{0\,\text{corrected}} = h_{0\,\text{radial}} \times \cos \phi \qquad\qquad (3.9)$$

3.5 The collector system

The impeller discharges into the collector system which is a static pressure recovery device. In most radial pumps this system is a volute (Figure 3.21(a)) and in some higher pressure rise machines a diffuser which is surrounded by a volute (Figure 3.21(b)). The vaneless diffuser (c) used in many compressors, is not applied to pumps due to its large size, and associated problems with design as a pressure casing, so the vaned diffuser is the design used. The two systems will be discussed in order.

3.5.1 The volute

This is a spiral casing (Figure 3.22) which slowly increases in area from the cut-water to the throat; the diffuser then connects the throat with the discharge flange. The cross-section of the volute may be circular, (as indicated in Figure 3.22) trapezoidal or rectangular in shape according to the practice followed by the designer. The flow

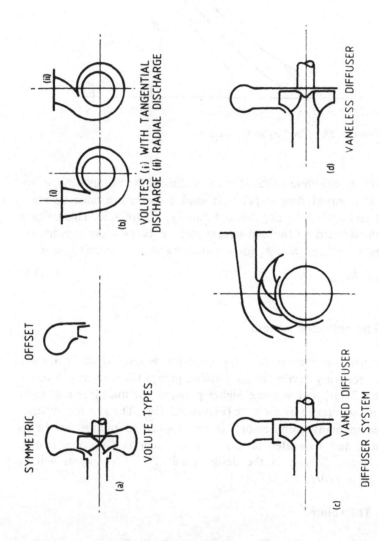

SYMMETRIC   OFFSET

(a)

VOLUTE TYPES

(b)

VOLUTES (i) WITH TANGENTIAL
DISCHARGE (ii) RADIAL DISCHARGE

(c)

VANED DIFFUSER

DIFFUSER SYSTEM

(d)

VANELESS DIFFUSER

Figure 3.21. Typical collector systems used in centrifugal pumps.

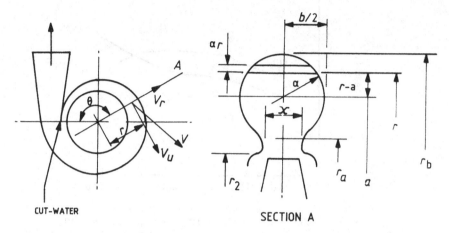

Figure 3.22.   A typical volute.

is collected in the spiral casing so that after 90° from the cut-water one quarter of the flow is assumed to pass through the cross-section, after 180°, half the flow, and so on. Two approaches to determining the area required are in common use, constant angular momentum, and constant velocity. The first method supposes that the tangential component of velocity multiplied by radius remains constant across any cross-section, so that integration of flow through an element as shown in Figure 3.22 is needed. The second assumes that the mean velocity in the throat is constant round the volute spiral. Figure 3.23 illustrates how the local pressure variation round the impeller periphery is affected by the two methods.

Of rather more importance is the choice of the correct volute throat area. Anderson (1980) proposed that the ratio of throat area to the impeller outlet area is one of the crucial factors in good pump design, and Worster (1963) presented a theoretical justification of the so-called Area Ratio in terms of a balance of volute resistance and impeller characteristic giving the design operating point (Figure 3.24). He also presented a 'design' plot, (Figure 3.25) indicating good practice using data from Anderson. Many designers now use this approach, as exemplified by Thorne (1979) at a recent conference on pump design.

Two alternative casing designs are illustrated in Figures 3.22 and 3.24. The first is a conventional 'close' casing where the height 'X' is slightly larger than the impeller passage height, and side-clearances

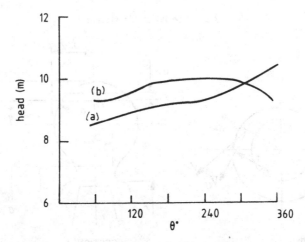

Figure 3.23. A comparison of the distribution of pressure around the periphery of a centrifugal pump due to a volute. (a) Designed on the assumption of constant angular momentum and (b) on the assumption of constant velocity distribution.

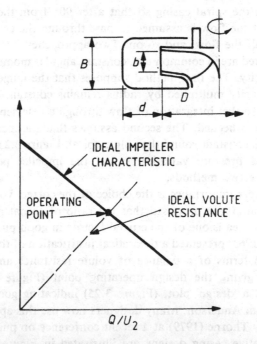

Figure 3.24. The definition of the best efficiency operating point using the intersection of the idealised volute and impeller characteristics (Worster (1963)). Courtesy of the Institution of Mechanical Engineers.

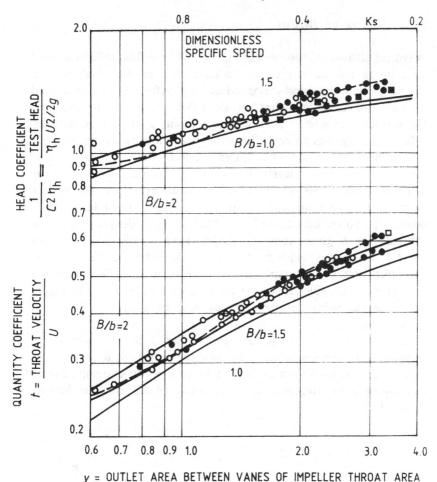

Figure 3.25. The Area Ratio design diagram (Worster (1963)).
Courtesy of the Institution of Mechanical Engineers.

are large enough to avoid manufacturing tolerance build-up giving impeller and casing interference. The second (Figure 3.24) illustrates the 'loose' casing favoured by some pump designers.

The shape of the cut-water nose is usually a radius, as 'sharp' as the casting process will allow, with the shape disposed on the notional stream-line followed by the liquid leaving the impeller. The method of establishing this geometry will be discussed in Chapter 6.

3.5.2 The vaned diffuser

A typical diffuser is illustrated in Figure 3.26. The fluid follows a spiral path up to the diffuser throat, is diffused over the length *bc* and discharged into the spiral casing. Ideally the diffuser vane walls should follow an Archimedian spiral but, since diffusers are often machined, the surfaces *abc* and *ef* are convenient radius approximations. Each diffuser passage has an active length *L*, which a working rule suggests can be between $3t$ and $4t$ to balance effective diffuser control and surface friction loss. Clearly the sum of throat areas will follow the same basic rule as the volute throat already discussed, and the number of vanes is always different from the number of rotor blades Z, a working rule often used being $Z + 1$. Many working designs that follow the rule for *L* given above appear to have a ratio for D_3/D_2 of 1.3–1.5, a constant width *b* which is a little wider than the impeller tip width and a maximum diffusion angle of 11°. The vanes are as thin as practicable, following the type of profile shown in Figure 3.26. Designs using aerofoil sections have been used, but are not conventional.

In multi-stage pumps the passages are partly diffusers and partly transfer ducts to ensure correct flow into the next stage impeller. Some typical passages are shown in Figure 3.27 for multi-stage pump designs; also shown is the effect on stage performance of the three duct layouts depicted.

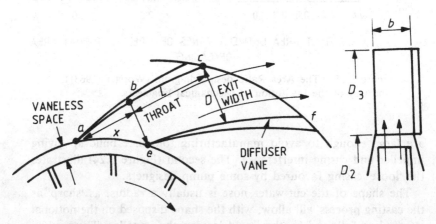

Figure 3.26. A typical vaned diffuser.

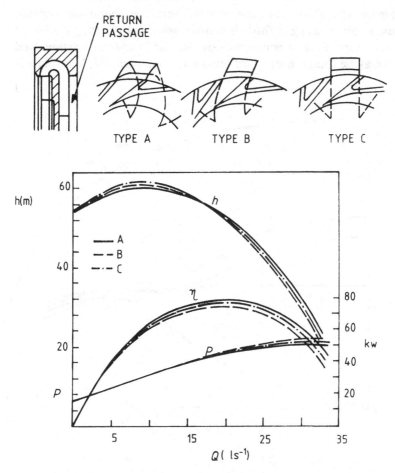

Figure 3.27. A comparison of the cross over duct designs that may be used to connect stages in a multi-stage pump, and their effect on the stage performance.

3.6 Thrust loads due to hydraulic effects

3.6.1 Radial thrust forces

In vaned diffuser pumps the pressure round the periphery of the impeller varies very little, so that any radial forces are very low, and do not vary with flow rate.

Figure 3.28 shows that the pressure variation round the impeller when a volute casing is fitted is considerable, and is strongly affected by flow rate. Several equations exist for thrust load estimation, and they may be found in standard textbooks. One such simple formula is

$$T_{\text{radial}} = kpD_2b_2 \qquad\qquad (3.10)$$

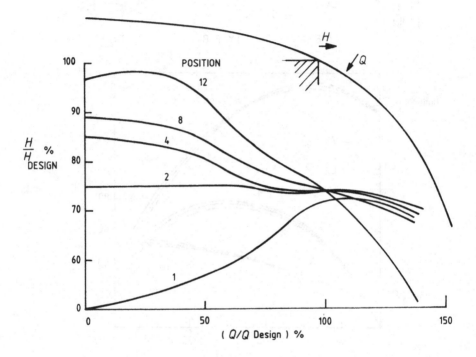

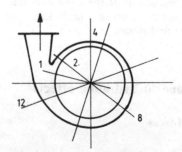

Figure 3.28. The variation in pressure around the periphery on an impeller at several different flow rates.

Figure 3.29 depicts a simple impeller and the quantities used in the equation.

Goulas and Truscott (1986) and Milne (1986) summarise the many contributions to the study of radial forces, and show that away from the design flow the radial force increases by a factor of up to 40:1 at low flows. The first mentioned demonstrates that there is a fluctuating component of about the same size as the 'steady' value for conventional pump designs. The designer must therefore size his shaft diameter and bearing system to deal with such forces, or provide a double volute casing so that opposing radial forces cancel each other out, but at the expense of extra flow loss.

3.6.2 Axial thrust forces

Axial hydraulic thrust is the summation of unbalanced forces acting on the impeller in the axial direction. Reliable thrust bearings are available so that this does not present problems except in large machines, but it is necessary to calculate the forces. These forces arise due to the distribution of pressure in the spaces between the impeller and the casing. In Figure 3.30 the assumption is illustrated for the discharge pressure acting over the backplate and front shroud in the single entry pump, and down to the wear rings for a double entry

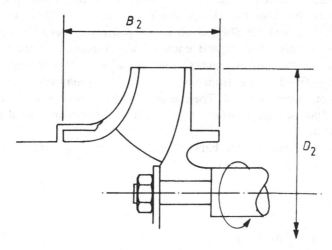

Figure 3.29. A definition of the dimensions used in Equation (3.10).

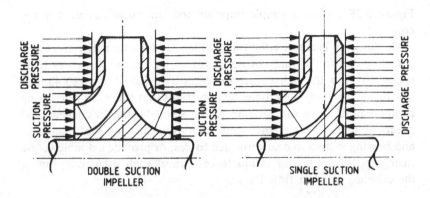

Figure 3.30. A simplified view of the axial pressure component and its distribution on the surfaces of single and double suction impellers.

impeller; the suction pressure is assumed to be distributed across the area up to the wear ring diameter in both designs. It can be argued that the pressure forces are balanced in the double suction design, and this is approximately the case in practice; the single suction impeller clearly has an unbalanced force. In practice there is always an unbalanced force acting on a double entry impeller due to such factors as unequal flow distributions in the two entry passages, and interference due to external features such as bends in the suction line, so that thrust bearings are always needed. Figure 3.30 illustrates the probable pressure distribution on the impeller surfaces. The actual pressure variation will depend upon surface roughness of the pump surfaces, on the side clearances, and on leakage flows through any wear rings fitted so that front and back net forces may vary from the design conditions assumed. The classic solutions assume that the fluid in the clearance space rotates at about half the impeller speed as a solid mass.

The axial thrust on the back plate due to pressure using a 'mean' pressure level p is given by

$$T_{\mathrm{BP}} = \frac{\pi}{4} \left[D_{\mathrm{w}}^2 - D_{\mathrm{s}}^2 \right] p \tag{3.11}$$

$$p = p_2 - \rho \frac{w^2}{8} \left[R_2^2 - r^2 \right] \tag{3.12}$$

Integration between D_{s} and D_{w} gives

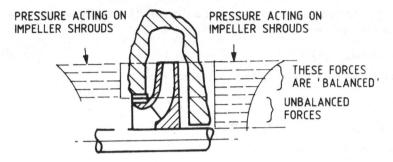

PRESSURE ACTING ON
IMPELLER SHROUDS

PRESSURE ACTING ON
IMPELLER SHROUDS

THESE FORCES
ARE 'BALANCED'

UNBALANCED
FORCES

Figure 3.31.   Pressure distribution and the free surfaces of a single
suction impeller accounting for the real flow effects of disc friction
and other flow effects.

$$T_{BP} = \frac{\pi}{4} [D_w^2 - D_s^2] \left[p_2 - \frac{1}{8} \left[u_2^2 - \left(\frac{u_w^2 + u_s^2}{2} \right) \right] \right] \tag{3.13}$$

p_2 is the *static* pressure at the impeller periphery.

The total pressure at the discharge flange p_3 is

$$p_3 = \rho g H - p_s \tag{3.14}$$

η_H is the *pump* hydraulic efficiency, and p_s is the suction pressure.

$$p_2 \simeq p_3 - \frac{V_T^2}{2} \tag{3.15}$$

On the reasonable assumption that volute flow losses approximate
to the regain in the diffuser, and the volute throat velocity V_T is the
volute kinetic energy.

The thrust on the impeller eye

$$T_s = p_s \frac{\pi D_w^2}{4} \tag{3.16}$$

Thus the net hydraulic thrust

$$T_1 = T_{BP} - T_s \tag{3.17}$$

T_1 acts as shown, and the momentum change in the axial direction
from the inlet duct into the impeller T_2 is given by

$$T_2 = \rho Q V_0 \tag{3.18}$$

So the net axial thrust $= T_1 - T_2$ $\tag{3.19}$

For an end suction design with an overhung impeller the net force on the shaft is

$$T_3 = (p_A - p_s)A_s \qquad\qquad (3.20)$$

Thus the net thrust is given by the equation

$$T = T_1 - T_2 \pm T_3 \qquad\qquad (3.21)$$

The sign of T_3 depends on whether

$$p_s < p_A \text{ or } p_s > p_A$$

The established texts may be consulted for other equations. The net result of the pressure forces and the fluid change of momentum is a force towards the suction flange; the force magnitude depends on size, outlet pressure, rotational speed and whether the impeller is provided with a shroud, as in many pumps, or is open. The most common way of reducing the axial thrust load is shown in Figure 3.32. A wear ring is formed on the impeller backplate, thus creating a chamber vented to the suction through 'balance holes'. Suction pressure is in this way applied to the backplate, up to the wear ring diameter, and hence reduces hydraulic force. This increases the risk of cavitation inception but is tolerated because it reduces the load. Other methods connect the balance chamber to the suction pipe by a balance pipe, or use pump-out vanes on the backplate of the impeller to create a radial variation in pressure in the space between impeller and casing.

Problems associated with multi-stage pumps have been discussed recently by Duncan (1986) and Ahmad *et al.* (1986), and they give much practical detail on correct design of the balance systems sketched in Figure 3.33.

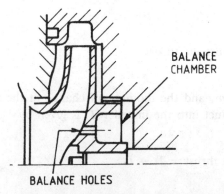

Figure 3.32. The balance chamber method of balancing axial thrust due to the pressure distribution.

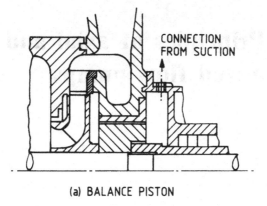

(a) BALANCE PISTON

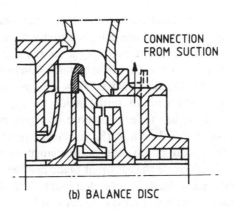

(b) BALANCE DISC

Figure 3.33.   Methods of balancing thrust in multi-stage pumps.

4 □ Principles of axial and mixed flow pumps

4.1 Introduction

Typical flow paths for axial and mixed flow machines are shown in Figure 4.1. Axial flow pumps are usually fitted with a rotor only, so that there is very little pressure recovery after the impeller and even where outlet guide vanes are fitted their main function is to remove any outlet swirl from the flow. Mixed flow pumps may either be as shown in Figure 4.1(b) without outlet guide vanes, or as in many machines, for example in the bulb or bore hole pumps, guide vanes are fitted to improve the flow into the second stage of the assembly.

Unlike the centrifugal pump, the performance in axial machines in particular is a function of the action of blade profiles. Only in mixed flow pumps with many blades is the dominant fluid dynamic action that of the passages as in centrifugal machines.

The fundamental relations have been introduced in Chapter 1, and the application of the Euler equation was demonstrated. In this chapter data for isolated aerofoils is discussed, as it applies to axial machinery, and the concepts of radial equilibrium and stall are introduced. This material forms the basis of empirical design techniques, where it is assumed that all stream surfaces are cylindrical. This is only approximately true in axial machines, and in mixed flow machines it is necessary to establish a number of stream surfaces and then either use the axial data along each surface, or use more advanced analytic fluid dynamic solutions based on the surfaces. This chapter therefore outlines an approach to stream surface shape determination and to mixed flow empirical and analytic solutions.

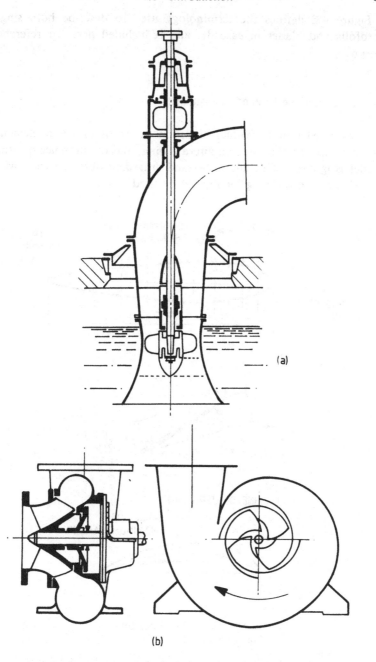

Figure 4.1.   Typical axial and mixed flow pump designs.

Figure 4.2 defines the terminology used to describe both single aerofoils and blades in cascade, and is included here for reference purposes.

4.2   The isolated aerofoil concept

It is assumed that individual blades in the rotor or stator elements interact with the fluid like an aircraft wing, and the presence of other blades is ignored. This allows lift and drag data obtained from wind tunnel research on wing sections to be used.

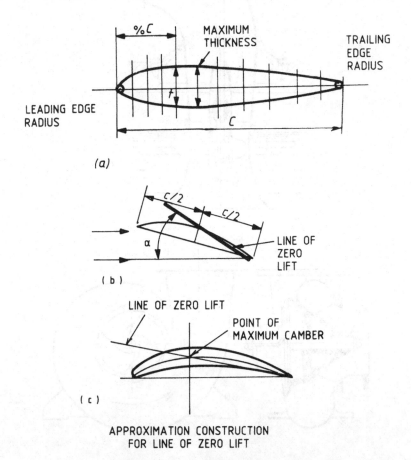

Figure 4.2.   Definition of single profile and cascade terms used.

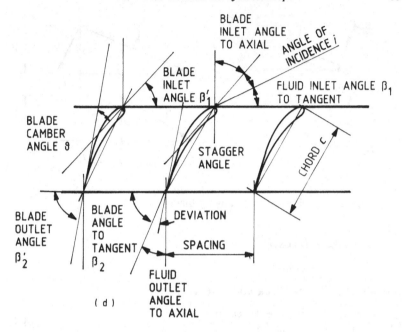

Figure 4.2.   *(cont.)*

4.2.1   Static blades

The lift force L is assumed to be at right angles and the drag D along the mean velocity direction. L and D are resolved in the axial and tangential directions (Figure 4.3) so that:

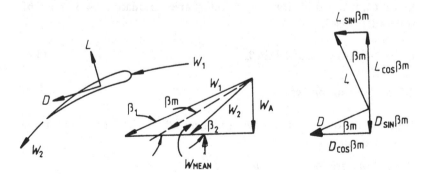

Figure 4.3.   Forces on a stationary blade profile.

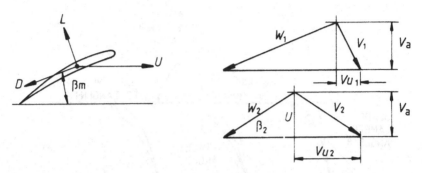

Figure 4.4.   Forces on a moving blade profile.

$$F_T = L \sin \beta_m + D \cos \beta_m$$

$$F_A = L \cos \beta_m - D \sin \beta_m$$

Introducing lift and drag coefficients C_L and C_D

$$F_T = \frac{\rho V_a^2 C}{2} \left[\frac{C_L \sin \beta_m + C_D \cos \beta_m}{\sin^2 \beta_m} \right] \qquad (4.1)$$

or

$$F_T = \rho V_a^2 s [\cot \beta_1 - \cot \beta_2] \qquad (4.2)$$

From these equations it can be shown that

$$C_L = 2 \frac{s}{c} [\cot \beta_1 - \cot \beta_2] \sin \beta_m - C_D \cot \beta_m \qquad (4.3)$$

Since $C_D / C_L = 0.05$ for many foil shapes, Equation (4.3) may be reduced to read:

$$C_L = 2 \frac{s}{c} [\cot \beta_1 - \cot \beta_2] \sin \beta_m \qquad (4.4)$$

4.2.2   Moving blades

Figure 4.4 shows a blade moving tangentially at velocity u, and in the process doing work.

Work done/second $= F_T \times u$

Conventionally a work (or head) coefficient $\psi = gH/u^2$ is used, and following the argument of Section 4.2.1, it can be shown that

$$\psi = \frac{V_a}{2u} \cdot C_L \cdot \frac{c}{s} \operatorname{cosec} \beta_m \left[1 + \frac{C_D}{C_L} \cot \beta_m \right] \tag{4.5}$$

If C_D/C_L is suppressed as being small

$$\psi = \frac{V_a}{2u} \cdot C_L \cdot \frac{c}{s} \operatorname{cosec} \beta_m \tag{4.6}$$

It may be noted in passing that if turbomachinery texts are consulted, the axial direction is often used as reference for β and then

$$\psi = \frac{V_a}{2u} \cdot C_L \cdot \frac{c}{s} \cdot \sec \beta_m \tag{4.7}$$

Thus if the ideal equations quoted in Chapter 1 are used to obtain the velocity triangles, β_m may be found, and if C_L is estimated, the c/s ratio determined as a next term. Real flow effects discussed later also need to be considered.

4.3 Blade data for axial flow machines

4.3.1 Isolated aerofoil information

Many 'families' of sections are available, and several publications present lift and drag data, typical of these being Abbot and Doenhoff (1959) for NACA data. Many texts also quote data for useful sections, as an example Figure 4.5 shows typical characteristics for four Gottingen profiles with 'infinite' aspect ratio. This type of data was obtained in a wind tunnel with Reynolds numbers based on chord in the region of 6–9 $\times$ 10^6, with low turbulence levels. Since the range of R_e in pumps and fans tends to be of the same order, the data is much used. Most of the profiles were designed as wings. The same criteria of minimum drag and control of separation to give an acceptable lift curve apply, even though the blades have a finite length.

It is not proposed to cover the theory of foils as this is covered in many textbooks, but the practical implications of Figure 4.6 taken from Carter (1961) will be discussed at this point. In general, blade profiles giving a 'concave' pressure distribution (Figure 4.6(a)) give a good low speed performance. In practical terms the choice of camber line and profile geometry determine blade performance. Carter discusses how a profile with the maximum camber well towards the leading edge gives a concave velocity distribution, a high stalling C_L, and a good

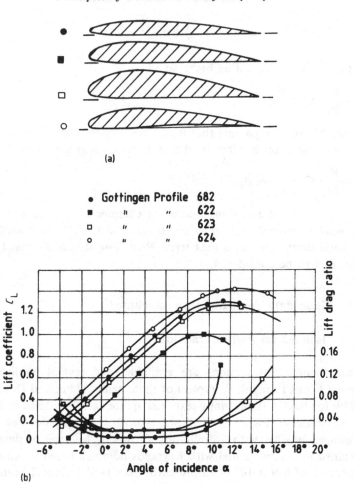

Figure 4.5.   Data for four different Gottingen profiles.

low speed performance but is prone to cavitation in pump applications. Moving the point of maximum camber back towards the trailing edges gives a lower stalling C_L, good cavitation behaviour, but a more limited operating range. The simplest circular arc camber line gives maximum camber at mid-chord, but Carter suggested a parabolic arc with maximum camber and 40% chord from the leading edge. Maximum thickness can be up to $12\frac{1}{2}$% chord for good performance; Table 4.1 gives the profile schedules of two common profiles, and

Table 4.1. *The standard profile dimensions for the NACA 65 010 and C4 profiles.*

% Chord	NACA 65,010 $t/2$	C4 $t/2$
0	0	
0.5	0.772	
0.75	0.932	
1.25	1.169	1.65
2.5	1.574	2.27
5.0	2.177	3.08
7.5	2.674	3.62
10	3.04	4.02
15	3.666	4.55
20	4.143	4.83
25	4.503	
30	4.76	5.0
35	4.9234	
40	4.996	4.89
45	4.963	
50	4.812	4.57
55	4.53	
60	4.146	4.05
65	3.682	
70	3.156	3.37
75	2.584	
80	1.987	2.54
85	1.385	
90	0.81	1.60
95	0.306	1.06
100	0	0
LE Radius	0.687% of chord	12% max t
TE Radius	'sharp'	6% max t

Figure 4.7 gives data for a typical profile, in this case Clark 'Y', which has a flat lower (suction) surface.

4.3.2   Cascade data

When blades are in close proximity, (the 'cascade' situation), the individual blade behaviour is affected, as Figure 4.8 indicates, since

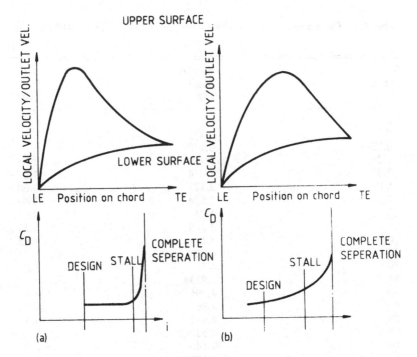

Figure 4.6. Typical drag coefficient variations for (a) a concave and (b) a convex pressure distribution. Based on Carter (1961). Courtesy of The Institution of Mechanical Engineers.

the passages formed by adjacent blades dominate the flow. A one-dimensional theoretical correction for this effect is to insert a correction factor in the Euler equation so that it reads, for a pump

$$gH = C_H (V u_2 u_2 - V u_1 u_1)$$ (4.8)

This factor is shown in Figure 4.9 and is a theoretical statement of the effect of blade spacing and blade angle. Weinig (1935) studied the two-dimensional problem by deriving relations for thin aerofoils which approximate to flat plates and produced a lattice coefficient K used to correct the flat plate equation

$$C_L = 2\pi K \sin \alpha$$ (4.9)

Figure 4.10 illustrates the effect on K of blade angle a and spacing.

The performance of cascades of blades was obtained using cascade tunnels, and the pressure and velocity changes at mid-height of the

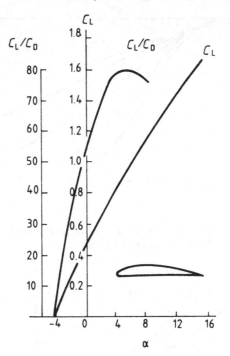

Figure 4.7.   Lift coefficient and lift to drag ratio variation with angle of attack for the Clark 'Y' profile.

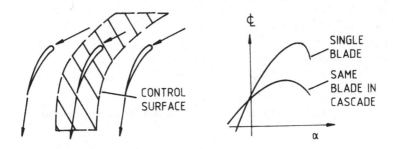

Figure 4.8.   A simplified cascade of blades.

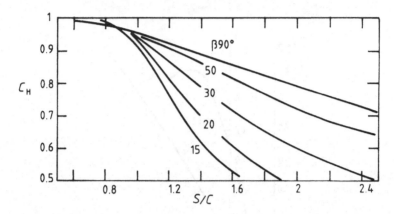

Figure 4.9. The relation of C_H with space chord ratio and setting angle for a simple plate cascade.

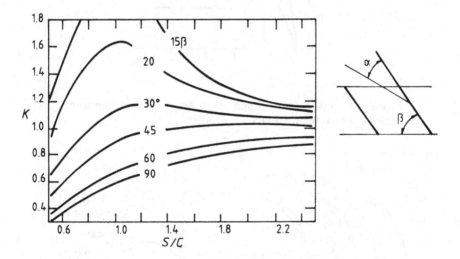

Figure 4.10. The Weinig lattice coefficient for plates in cascade.

centre blade and its associated passages found by measurement. The tunnels had variable geometry to produce variations in incidence, and wall boundary layer control to ensure a two-dimensional flow field at mid-height for the section under study. Figure 4.11 illustrates the typical performance of a cascade of NACA 65 series blades. The design lift coefficient C_{L0} is related to camber for this foil shape

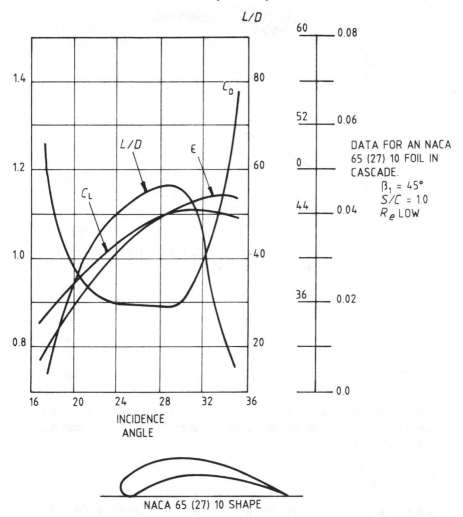

Figure 4.11. The performance of an NACA series 65(27)10 profile in cascade.

in the way shown in Figure 4.12. Texts such as Horlock (1958) Turton (1984) and Dixon (1975) discuss the range of data for compressors, fan, and turbine blading, but it is necessary here to introduce some data published by Howell (1945). He introduced a diffuser or row efficiency

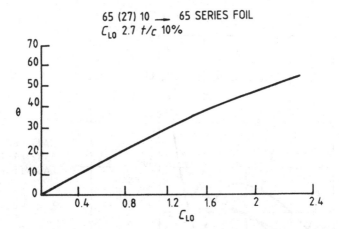

Figure 4.12. The variation of the design lift coefficient with camber angle for an NACA series 65(27)10 profile.

$$\eta_D = \frac{p_2 - p_1}{\frac{1}{2}\rho(v_1^2 - v_2^2)} \qquad (4.10)$$

and it can be shown that

$$\eta_D = 1 - 2\frac{C_D}{C_L}\cos 2\beta_m \qquad (4.11)$$

Howell also showed (assuming C_L/C_D is constant, which is approximately true for many profiles over a range of incidence) that β_m optimum is 45° and that

$$\eta_D = 1 - \frac{2C_D}{C_L} \qquad (4.12)$$

suggesting η_D is maximum for a compressor row where β_m is 45°. Howell illustrates how η_D varies with β_m and the lift/drag ratio and demonstrates the small effect of variation in C_L/C_D for a conventional cascade.

Cascade testing has revealed strong two- and three-dimensional flow patterns, the principal effects being illustrated in Figure 4.13. So boundary layer and tip clearance effects have a strong effect on the overall efficiency of a blade row, and, due to the secondary and wake flows affecting blades in succeeding rows, the performance of the whole machine.

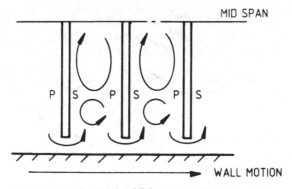

(a) EFFECT OF RELATIVE MOTION, WALLS AND BLADES

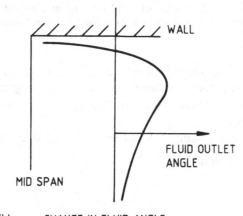

(b) CHANGE IN FLUID ANGLE
ALONG BLADE DUE TO WALL B.L.

Figure 4.13. Three-dimensional flow effects in an axial cascade of blades.

4.3.3 Radial equilibrium concepts

The idealised theory assumes that the stream surfaces are all cylinders. The action of the blading, however, creates a shift of particle tracks. A simple approach much used is to assume 'radial equilibrium' in the flow before and after the blade row, with radial adjustment in position taking place through the row. Figure 4.14 illustrates a small fluid element rotating in a pressure field at constant tangential velocity

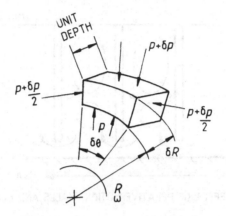

Figure 4.14.   A particle of fluid rotating at constant rotational speed and radius and the forces needed for equilibrium.

V_u ($V_u = \omega R$). Applying Newton's laws, the particle can only move in a circular path radius R if there is a radial pressure gradient providing the necessary centripetal force. If the field is such that the pressure gradient is dp/dR.

If the element volume is $R\,dR$ and $d\phi$, and fluid density is

$$R\,dp\,d\phi = \rho V_u^2 dR\,d\phi \tag{4.13}$$

or

$$\frac{dp}{dR} = \frac{\rho V_{u2}}{R} \tag{4.14}$$

or

$$\frac{dP}{\rho} = V_u^2 \frac{dR}{R} \tag{4.15}$$

If we bring in the concept of stagnation pressure

$$p_0 = p + \tfrac{1}{2}\rho V^2,$$

$$\frac{1}{\rho}\frac{dp_0}{dR} = V_a \frac{d(V_a)}{dR} + \frac{V_u}{R}\frac{d(RV_u)}{dR} \tag{4.16}$$

The radial equilibrium concept assumes p_0 does not vary with R,

therefore $V_a \dfrac{d(V_a)}{dR} + \dfrac{V_u}{R}\dfrac{d(RV_u)}{dR} = 0$

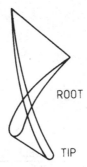

Figure 4.15. The relation between the root and tip sections of an Axial blade designed on free vortex principles.

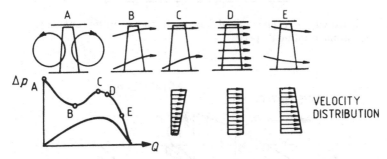

Figure 4.16. A description of the likely flow patterns to be found in an axial flow machine at several flow rates.

or

$$\frac{d(V_a)^2}{dR} + \frac{1}{R}\frac{d(RV_u)^2}{dR} = 0 \tag{4.17}$$

A solution of these equations that is much used is the simplest model (the free vortex concept):

$$V_u \times R = \text{constant} \tag{4.18}$$

Figure 4.15 illustrates the effect on a blade length of this concept.

This concept works well at design point, but as Figure 4.16 shows, away from this flow very complicated flow patterns may occur.

4.3.4 Stall effects

In axial machines particularly, all or some of the blades stall due to local flow effects, and the resulting machine characteristic, Figure 4.17 indicates flow instability.

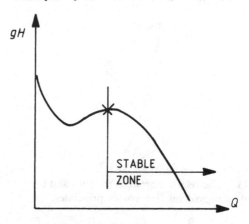

Figure 4.17.   A typical axial flow pump characteristic running at constant rotational speed.

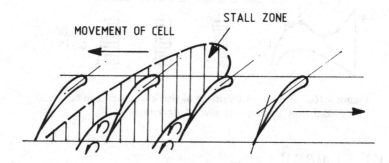

Figure 4.18.   Rotating stall in an axial flow blade row.

Below the stall point, flow is unpredictable, with the line being a 'best fit', and some 'surging' in motor current can occur, with noise. In some cases the effect can be rotating stall, Figure 4.18, which has been studied exhaustively in fans and compressors by Smith and Fletcher (1954) Dunham (1965) Lakhwani and Marsh (1973) among others.

4.4   An approach to mixed flow machines

Where pumps are of radial, 'Francis' type or completely mixed flow layout, the principles for centrifugal pumps already covered are often used, once the stream surfaces are established. In the following sections the approach to the shape of stream surfaces is discussed, and then the empirical and computer based approaches are outlined.

4.4.1   Stream surface design

Solutions are described in some detail in texts such as that by Wisclicenus (1965) and in less detail by Turton (1984). An approximate solution to the three-dimensional flow problems involved for a curved passage is illustrated in Figure 4.19. If a rotational flow is passing through the passage, but no blades are present, and it is assumed

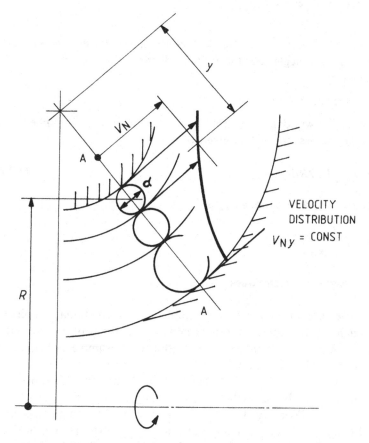

Figure 4.19.   Idealised flow in a curved flow passage.

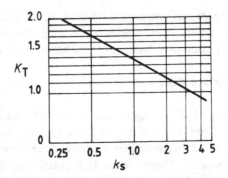

Figure 4.20. Variation of the axial thrust load factor K_T with the characteristic k_S.

that the free vortex law applies with constant angular momentum, and flow losses are neglected, the velocity distribution along line AA will be given by

$$V_N y = \text{Constant} \tag{4.19}$$

The constant will depend on the flow field if continuity is applied, the flow rate across line AA is given by

$$Q = \sum_{AA} V_N 2Rd \tag{4.20}$$

So that if V_N at a distance y, is V_{Ny}

$$Q = \sum \left(\frac{V_{Ny} y_1}{y} \right) 2\pi Rd \tag{4.21}$$

4.4.1 Empirical techniques

Once the stream surfaces are determined, conventional empirical approaches use blade profile data and lay out the profiles along the surfaces in mixed flow machines, as outlined for example by Stepanoff (1976) in his design charts.

A technique which is mainly empirical and has been developed for computer use is that described by NEL designers for mixed flow fans and pumps. Wilson (1963) describes the design of a pump of specific speed (English) 5900, and later contributions by Myles (1965), Stirling (1982) and Stirling and Wilson (1983) show how the technique has developed to provide designs.

4.5 Thrust loads

4.5.1 Axial machines

Since the flow is axial, any radial loads can only arise due to out of balance. Axial thrust arises due to the axial load on the blades discussed in Section 4.2.2, and to the net load applied to the impeller disc due to the pressure difference across the impeller. The thrust load due to pressure difference across the impeller vanes can be calculated from

$$T_{\text{Axial}} = A\rho \frac{gH}{\eta_{\text{H}}} \tag{4.22}$$

where A is the annulus area swept by the blades, and gH is the actual total specific energy rise across the impeller. To this must be added

$$T = A_{\text{D}}\rho \, gH_{\text{Static}} \tag{4.23}$$

A_{D} is the impeller disc area acted upon by the static pressure rise.

4.5.2 Mixed flow machines

An approximate formula quoted by Stepanoff and others is

$$T = A\rho gHK_{\text{T}} \tag{4.24}$$

where A is the inlet annulus area, and K_{T} is a factor shown in Figure 4.20.

5 □ Flow calculations in pumps and an introduction to computer aided techniques

5.1 Introduction

In the study of fluid mechanics, Newton's second law of motion enables the Eulerian equations of motion of a fluid to be applied to a study of the forces acting on a fluid particle at a particular point at a particular time. The solution of these equations with the true boundary conditions in a pump is a formidable task, because within a pump there are rotating and stationary blades that change in their orientation and cross-sectional geometry from hub to tip. There also are boundary layers on the annulus walls and the blade surfaces, wakes from the trailing edges of the blade, over-tip leakage flows etc, so the flow is unsteady, three-dimensional, and viscous.

A brief reference has already been made to empirical approaches in the chapter on axial and mixed flow machines principles. In this chapter the empirical approach to determining passage shapes will first be outlined, and then the more analytical techniques made possible by the computer will be outlined.

5.2 Stream-surfaces

Where pumps are of radial, of Francis type or completely mixed flow layout, the principles for centrifugal pumps already covered are often used, once the stream-surfaces are established. In the following sections the approach to the shape of stream surfaces is discussed, and then empirical solutions are outlined.

5.2.1   Stream-surface design

Solutions are described in some detail in texts such as that by
Wisclicenus (1965) and in less detail by Turton (1984a). An approximate solution to the three-dimensional flow problems involved for a
curved passage is illustrated in Figure 5.1. If a rotational flow is passing
through the passage, but no blades are present, it is assumed that the
free vortex law applies with constant angular momentum, and flow
losses are neglected, the velocity distribution along line XX will be
given by

$$V_{Ny} = \text{constant} \tag{5.1}$$

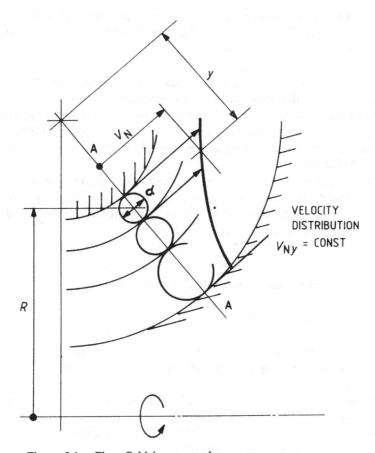

Figure 5.1.   Flow field in a curved passage.

The constant will depend on the flow field. If continuity is applied, the flow rate across line XX is given by

$$Q = \sum_{XX} V_N 2Rd \tag{5.2}$$

So that if V_N at a distance y_1 is V_{Ny}

$$Q = \frac{(V_{Ny}y_1)}{y} 2Rd \tag{5.3}$$

5.3 Empirical techniques

Once the stream-surfaces are determined conventional empirical approaches use blade profile data and lay out the profiles along the surfaces in mixed flow machines, as outlined for example by Stepanoff (1976) in his design charts.

A technique which is mainly empirical and has been developed for computer use is that described by NEL designers for mixed flow fans and pumps. Wilson (1963) describes the design of a pump of specific speed (English) 5900, and later contributions are by Myles (1965).

Stirling (1982) and Stirling and Wilson (1983) show how the technique has developed to provide good pump designs.

The empirical techniques described have been computerised as is outlined by the later contributions noted in the previous paragraph, and a number of such approaches are summarised in a recent study performed by Hughes *et al.* (1988).

5.4 Computer based theoretical techniques

5.4.1 The basic equations

The laws of conservation of mass, momentum and energy together with some equation relating pressure to density and temperature form a system of six partial differential equations with six unknowns: the three components of velocity, pressure, density and energy. The averaging of the equations to remove the time dependent terms from the equation leads to time averaged correlations of the turbulence components which add to the complexity of the equations.

In a vector form the equations are given as:

$$\nabla(\rho v) = 0.$$

$$\mathbf{w} \cdot \nabla w + 2\omega \cdot w - \omega^2 \mathbf{R} = -\frac{1}{\rho} \nabla p + \mathbf{F} \tag{5.4}$$

$$\nabla \cdot (\rho \mathbf{W} I) = \nabla \cdot (k_{\text{Eff}} \nabla T) + \nabla \cdot \tau \mathbf{w} \tag{5.5}$$

where I is the rothalphy defined as:

$$I = H_0 - \omega R V_u$$

and τ is a stress tensor.

F is a vector which contains gradients of Reynolds and viscous stresses. To obtain solutions, assumptions are needed related either to the flow itself or to the geometry.

5.4.2 Inviscid cascade flow

In this case the force F is zero. If the density is constant the equations of motion become the Euler equations. The first important corrollary of Euler's equations is Kelvin's circulation theorem which states that the circulation around a loop consisting continuously of the same fluid particles is conserved, i.e.

$$\frac{\Phi\omega}{Dt} \int V \cdot dl = 0 \tag{5.6}$$

We may now introduce the idea of vorticity which is defined as:

$$\omega = \nabla \times v \tag{5.7}$$

This has the property that if

$$\omega = 0, \quad \text{then} \frac{D\omega}{dt} = 0$$

If a fluid particle has no vorticity at some instant then it can never acquire any.

This result is known as the permanence of irrotational motion. It corresponds physically to the fact that, when Euler's equation applies, the only stresses acting on a fluid particle are the pressure stresses. These act normal to the particle surface and so cannot apply a couple to the particle to bring it into rotation.

In irrotational motion

$$\omega = 0$$

throughout the flow. Hence, one may introduce ϕ such that

$$V = \text{grad } \phi \tag{5.8}$$

ϕ is known as the velocity potential (by analogy with other potentials).
 From continuity

$$\text{div } V = 0$$

and so

$$\nabla^2 \phi = 0 \tag{5.9}$$

the velocity potential obeys Laplace's equation, which can be solved by
mathematical means for two- or three-dimensional flows. At present
the equations are approached using either stream-line curvature or
stream-surface solution techniques, and these will now be discussed.

5.4.3 Stream-line curvature method

This was developed over 20 years ago, and discussed initially in, for
example, Hamrick *et al.* (1952), Wood and Marlow (1966) and more
recently by Casey and Roth (1984).

 The method considers the equations governing an element of fluid
at mid-point P of a tube of fluid spanning the gap between the adjacent
blade surfaces. Figure 5.2 shows strongly tilting blades typical of a
mixed flow pump. P lies on a meridional stream-line, (Figures 5.3
and 5.4) and the local velocity and acceleration components are related
to forces caused by the blade surfaces, and the accuracy of modelling

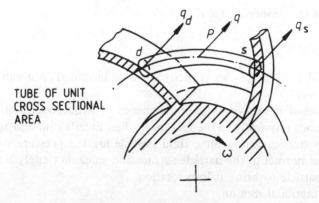

Figure 5.2. Flow between adjacent vanes.

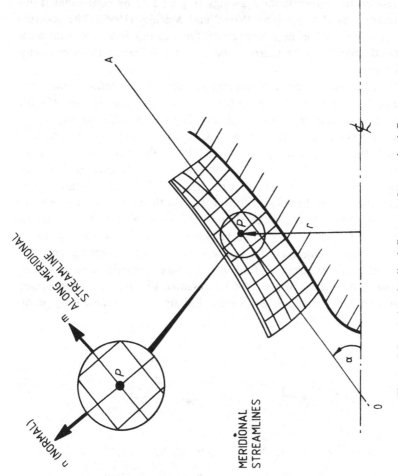

Figure 5.3. A longitudinal flow net for a mixed flow pump

is greater as blade numbers increase. The shape of the stream-surfaces is related to conventional drawings (Figure 5.5) by conformal transformation as described by Wood and Marlow (1966). The solution when refined using the approach outlined in this reference relates the rate of change of pressure along the streamline to the stream-line curvature.

Solution for a series of stream-surfaces along defined lines is a predictive technique. For example, consider a typical impeller (Figure 5.6). Speed, flow rate and geometry are fed in, solution starts at a known point on the hub say, with an assumed velocity, and proceeds across the flow, settling streamline intervals that enclose a fixed percentage of the flow, say 10%, terminating at the shroud. The total flow is integrated, compared with that given, and the iteration continues till the two flow rates converge. This then repeats for successive hub positions until the entire flow field is described, with mid-channel velocities found. Strictly these velocities occur on the vane surfaces, but as Wood suggests the concept of constant rotating enthalpy (rothalpy) can be used to find vane surface velocities and pressures. In the lecture series Wood (1986) presented two pump solutions, and Figures 5.7, 5.8 and 5.9. Both were designed for the same duty point,

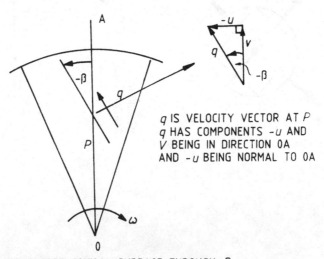

q IS VELOCITY VECTOR AT P
q HAS COMPONENTS $-u$ AND
V BEING IN DIRECTION 0A
AND $-u$ BEING NORMAL TO 0A

DEVELOPED CONICAL SURFACE THROUGH P

Figure 5.4. A statement of velocity components at a point in a flow passage.

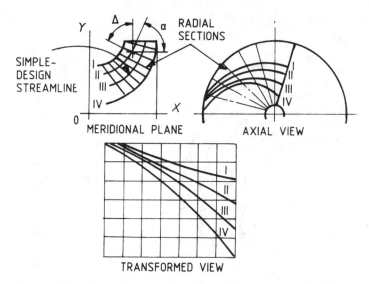

Figure 5.5. Basic diagrams for the traditional method of blade and passage design. (After Wood and Marlow (1966)).

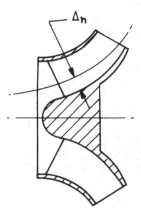

Figure 5.6. A typical mixed flow impeller.

Type 2 peaked at 89%, Type 1 at 83%. It may be commented that: the streamlines for Type 2 are more regularly spaced and follow the hub and shroud curvature more closely than those for Type 1. Type 2 does not show such pronounced velocity maxima downstream of the leading edge. It is considered to be desirable for the mid-channel

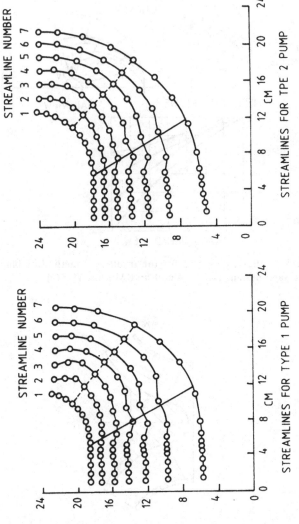

Figure 5.7. Stream surfaces for two pump designs. (After Wood (1986)).

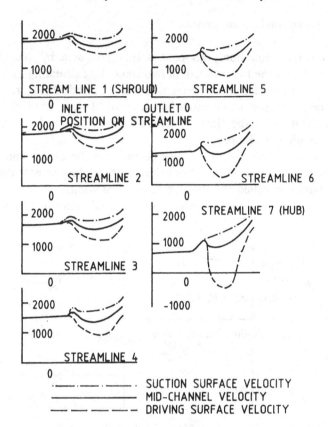

Figure 5.8.   Velocity distributions for pump Type 1. (After Wood (1986)).

velocities to increase progressively towards the trailing edge (Type 2 is superior in this respect to Type 1). The blade loading, which is characterised by the difference between velocities on the suction and driving surfaces, should increase gently from the leading edge, hold nearly constant over the middle of the blade length and decrease towards the outlet. Again, Type 2 is superior to Type 1 in this respect.

These comparisons are qualitative and retrospective but nevertheless give designers a guide towards the desirable features of a satisfactory hydraulic design.

5.4.4 Stream-surface techniques

The philosophy commonly adopted is that proposed by Wu (1952) namely to tackle the full problem in two stages by calculating the flows on two intersecting families of stream-surfaces S1 and S2 (Figure 5.10). The S1 family of surfaces is essentially a set of blade-to-blade surfaces and the solution of the flow on each of these surfaces is called the blade-to-blade calculation.

The S2 family of stream-surfaces lie between the blades and span the space from hub to shroud. The solution of the flow on each surface of this family is called the through-flow calculation.

Using Wu's general theory the complete three-dimensional flow field

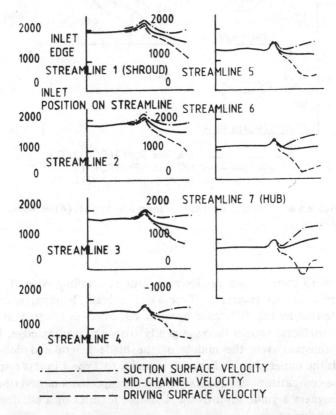

Figure 5.9.   Velocity distributions for pump Type 2. (After Wood (1986)).

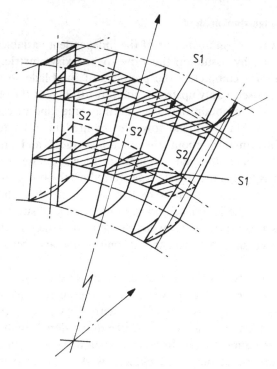

Figure 5.10.   The S1 and S2 stream-surfaces proposed by Wu (1952).

through a blade row could be obtained by adopting an iterative procedure to link the calculations for the flow on the two families of stream-surfaces. However, in general this procedure has not been adopted. The majority of workers have used the philosophy of Wu's S1 and S2 stream-surfaces without adopting an iterative link. The information required to pass from one type of stream-surface to another is the geometry of the stream-surface and the stream-sheet thickness.

One commonly adopted procedure is to use only one S2 stream-surface through the blade row together with a number of S1 stream-surfaces; with this approach the stream-surfaces must by necessity be axisymmetric.

The main advantage of such a scheme is the elimination of one of the independent variables thus reducing the problem to a two-dimensional one. This allows the use of a stream function formulation to achieve a solution of the flow on the surface.

5.4.4.1 Through-flow method

One of the ways to eliminate one of the independent variables, namely
θ, can be done by assuming the flow to be axisymmetric. Such an
approach ignores completely the presence of the blade so the information will be useful only upstream and downstream of the blade. This
can also be used to obtain the losses during some kind of correlation.

If stations within the blade row are required in order to provide
additional information about the stream-surfaces and stream-tube
heights for the blade-to-blade calculation, then it cannot be assumed
that the flow is axisymmetric. In order to reduce the problem to a two-
dimensional one information is needed about the three-dimensional
blade shape and the loss distribution along the blade surfaces. In the
initial stages this will not be available and consequently there must
be an iterative link between the through-flow and blade-to-blade
calculations.

As has been noted already, when only one S2 through-flow surface
is used to define the S1 stream-surfaces for the blade-to-blade calculation these have to be taken as axisymmetric. As a consequence this
approach is referred to as a quasi-three-dimensional analysis rather
than fully three-dimensional. In practice even if the S1 stream-surfaces
were axisymmetric at inlet to a blade row they would twist as they
passed through the blade row under the influence of streamwise or
secondary vorticity. The magnitude of the secondary flow (i.e. deviation away from axisymmetric flow) is governed among other things
by the inlet conditions to the blade row, blade forces, blade rotation
etc. The effect of the blade can be taken into account using one of
the following three approaches:

(a) Distributed body force. Here the flow is taken as axisymmetric with
     a distributed body force. This force is related to a force normal to
     a mean stream-surface or blade camber line, and can be related to
     the pressure gradient at each point between the pressure and suction
     sides of the blade.
(b) The mean S2 stream-sheet approach. In this the flow is assumed to
     follow a mean S2 stream-sheet and as a consequence a force normal
     to the stream-sheet occurs in the equations of motion. This allows
     the replacement of one of the equations of motion with geometric
     conditions relating the three components of velocity. There is a
     certain amount of arbitrariness in the mean S2 stream-sheet that
     should be adopted. In certain approaches the blade mean camber line

has been used. Other approaches have used a stream-sheet determined iteratively. For example, from a through-flow analysis using
an approximation to the mean stream-sheet a series of blade-to-blade
calculations can be performed at a number of radial sections; the
mean stream-sheet definition can then be up-dated and the procedure
repeated.

(c)   The blade passage-averaging technique. Here the equations of motion
are integrated in the circumferential direction between one blade
and the next. In this way blade passage-averaged equations result
which include the effects of the blade blockage and blade force.
This removes the degree of arbitrariness apparent in (b) above and
allows one to see more clearly the approximations being employed.

5.4.4.2   Blade-to-blade methods

In the case of blade-to-blade stream-surfaces the main variables are
R–θ or Z–θ. Averaging of the quantities is not possible so the geometry
of each surface must be the result of a through flow calculation.
Furthermore an extra force must be included to ensure that the flow
follows the stream-surface. The added complexity on a blade-to-blade
stream-surface comes from two sources: the need to apply the
periodicity condition upstream and downstream of the blade, and the
treatment of the flow around the leading and trailing edges where
strong changes in curvature may occur.

5.4.5   Methods of solution

Having formulated the problem, which is now reduced to a two variable
problem on a stream-surface, a solution of the equations of continuity,
momentum aud energy must be sought.

The most common approach is to use a stream function definition
in the form:

$$\frac{\partial \psi}{\partial x} = \rho B W_y$$

$$\frac{\partial \psi}{\partial y} = -\rho B W_x$$

which will satisfy exactly the continuity equation. The momentum
equation must then be solved to obtain the distribution of the stream
function and, by differentiating, the velocity distribution. Two of the
methods used to solve the above equations are the 'matrix through flow
analysis' and the 'stream-line curvature' method, and work has been
done using finite element techniques and computer solution. A detailed

discussion of the techniques is not possible in this treatment, so reference may be made to such courses as that offered at Cranfield (Wood, 1986), Von Karman Institute lecture series and the many working papers cited in such proceedings.

A considerable number of approaches have been developed for centrifugal and axial compressors, which have a high added value and thus investment has been justified. A recent study in which the author was involved (Hughes *et al.*, 1988) summarised the situation, and it was concluded that very little original software has been produced for pumps, and much of that which is available is basically for compressors with substituted sub-routines to deal particularly with cavitation and liquid properties. The study cites both commercial packages and academic papers, and forms a useful additional source of reference, with a large section on the interfacing of design with manufacture.

A further source of information on the semi-empirical computer solutions developed by the National Engineering Laboratory will be found in the lecture series on pump design, NEL (1983). The approaches are clearly presented, and the resulting designs for radial, mixed flow and axial machines are compared with experimental data.

6 □ Single stage centrifugal pump design

6.1 Introduction

When designing an impeller the suction diameter, outlet diameter, and leading proportions of the blade passages need to be determined before the blade profiles are determined.

Since $NPSH_{required}$ is an increasingly important criterion in process pump application, the suction area design is crucial, and will be discussed first. The determination of the outside diameter, outlet angle and the number of blades is then described. The design of passage and blade profiles is then examined.

6.2 Initial calculations

The design steps will be illustrated using a worked example. The duty to be achieved is $250 \, m^3 \, hr^{-1}$ of water. (1101 US GPM) at a head of $55 \, m$ (180.5 ft) with an $NPSH_{available}$ $2.5 \, m$ (8.2 ft).

The rotational speed should be decided first. As a first guess such aids as the Hydraulic Institute Standards (1983) can be consulted, but company experience suggests 3000 rpm to give a smaller size, though 1500 rpm would be chosen by using the standard. Using 2950 rpm, (allowing for motor slip)

$$k = \frac{2950\sqrt{0.069}}{(g \times 55)^{3/4}} = 0.725 \tag{6.1}$$

from Figure 1.8. $\eta = 82.5\%$

95

Thus, since design water power $= 10^3 \times g \times 55 \times 0.069$ kW
probable shaft power $= 45.13$ kW.
from Figure 2.10 $\sigma = 0.2$ so $\text{NPSH}_R = 11$ m and $p_S = 1.08 \times 10^5$ N m^{-2}.

Referring to Figure 3.13, the impeller should have a mixed flow inlet
and a parallel axisymmetric outlet, and the characteristic should rise
to shut valve flow, with a diameter ratio of between 1.5 and 2.

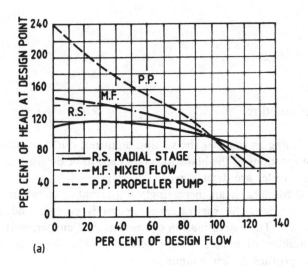

(a)

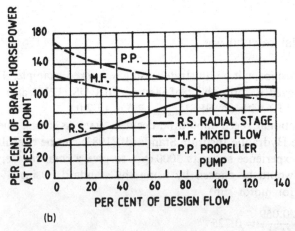

(b)

Figure 6.1. Empirical charts relating head, power and efficiency
to flow rate.

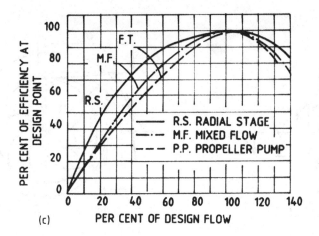

Figure 6.1. (*cont.*)

In many cases the customer will require a statement of the 'shut-off' head, and an estimate of the probable flow rate to head curve shape and also some idea of the power consumption at low flows and at flow rates higher than design. An empirical method using the curves shown in Figure 6.1 suggests that for this machine the shut valve head and power will be 75 m and 27 kW respectively, and the corresponding values at 130% of the duty flow will be 36 m and 43 kW.

Recently there has been an attempt to provide a more accurate method of finding the shut-off head published by Frost and Nilsen (1991). They surveyed the available data and formulae, and came to the conclusion that the shut valve head is relatively insensitive to the number of blades, the blade outlet geometry, and the channel width at outlet from the impeller. They outline a theoretical approach to the problem, based on Figure 6.2, using the concept that the Head at shut valve H_{SV} is found from

$$H_{SV} = \text{Impeller head}_{SV} + \text{Volute head}_{SV} \tag{6.2}$$

$$(\text{or } H_{SV} = H_{\text{imp SV}} + H_{\text{vol SV}})$$

where

$$H_{\text{imp SV}} = \frac{\omega^2 R_2^2}{2g} \left[1 - (R_S/R_2)^2 \right]$$

and

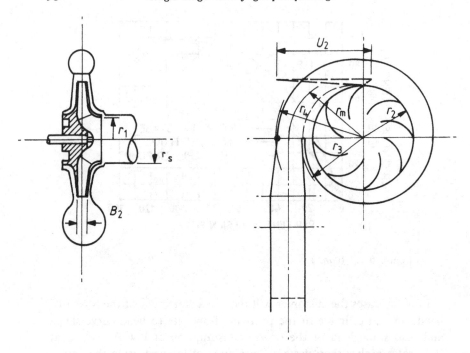

Figure 6.2.   A typical centrifugal pump using dimensions defined by Frost and Nilsen (1991). (With acknowledgement to the Institution of Mechanical Engineers).

$$H_{\text{vol SV}} = \left(\frac{\omega R_2}{R_M - R_2}\right)^2 \left[R_M^2 \ln R_4/R_2 - 2R_M(R_4 - R_2) + \frac{(R_4^2 - R_2^2)}{2}\right] 1/g$$

The authors comment that the error in using this formula compared with that quoted by Worster (1963) is +2.5% and when used with data for pumps from two manufacturers averaged $-3.6 \pm 2\%$ when compared with actual test from points.

When shut-off head is calculated the value will be found from the dimensions, when established, using Equation (6.2) and compared with the values found from Figure 6.1.

6.3   Suction geometry

Using the approach outlined in Section 2.5,

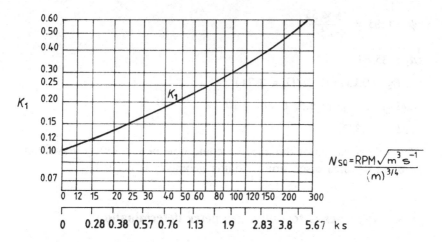

Figure 6.3. A plot of the design coefficient K_1 against k_S. (Based on Stepannof (1976)).

$$D_S = 4.66 \left(\frac{Q}{N} \right)^{1/3} = 0.134 \, \text{m}$$

An alternative approach is the empirical method presented by Stepannof (1976), Figure 6.3 being based on his data. $k = 0.725$

$K_1 = 0.185 \therefore V_{N_1} = 0.185 \sqrt{2g55} = 6.1 \, \text{ms}^{-1}$.

Suction velocity empirically lies between $(0.9$ and $1.0) \times V_{N_1}$, thus $V_S = 5.47$

$\therefore D_0 = 0.127$ – we will use $0.125 \, \text{m}$.

Clearly experience of similar pumps will guide the designer in the choice of the suction diameter. The value of 0.125 will be used rather than the figure of 0.134 using the approach of Section 2.5, as the risks of recirculation have already been outlined in Section 3.3.

The correct outlet geometry must now be determined.

6.4 First calculation of flow path

The outside diameter needs to be tentatively determined, using Figure 6.4.

$$\psi = 0.85 = \frac{2g \times 55}{u_2^2}$$

$u_2 = 35.63$

$\therefore D_2 = 0.231\,\text{m}\,(D_2/D_0 = 1.71)$

and from Figure 6.5, $D_2/b_2 = 10$.

$\therefore b_2 = 0.023$.

The side elevation of the impeller could now be tentatively drawn but a more detailed discussion of outlet geometry is needed.

6.5 Approaches to outlet geometry determination

Several empirically based approaches to outlet geometry are available, the slip concept may be used, or the area ratio method, or design

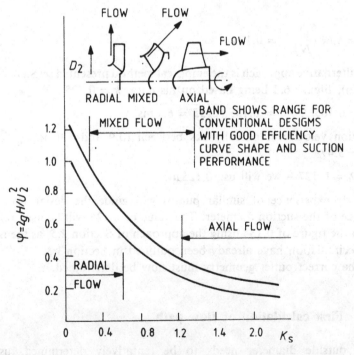

Figure 6.4. A plot of specific energy coefficient against k_s.

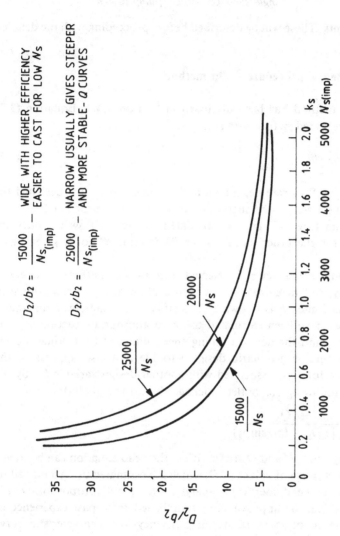

Figure 6.5. An empirical chart relating outlet geometry to k_S.

$$D_2/b_2 = \frac{15000}{N_{S(imp)}} \quad - \text{WIDE WITH HIGHER EFFICIENCY} \\ \quad - \text{EASIER TO CAST FOR LOW } N_S$$

$$D_2/b_2 = \frac{25000}{N_{S(imp)}} \quad - \text{NARROW USUALLY GIVES STEEPER} \\ \quad - \text{AND MORE STABLE-} Q \text{ CURVES}$$

$\dfrac{25000}{N_S}$

$\dfrac{20000}{N_S}$

$\dfrac{15000}{N_S}$

D_2/b_2

k_S

$N_{S(imp)}$

coefficients. These will be described before proceeding with the detailed design.

6.5.1   Design procedure – slip method

The slip method has been discussed in Section 3.4.4. Equation (3.7) will be re-stated before using it.

$$gH = \eta_h u_2^2 \left[h_0 - \frac{V_{m_2}}{u_2 \tan \beta_2} \right] \tag{3.7}$$

Since V_{m_2} is flow rate dependent and the impeller inlet is designed for optimum efficiency and suction performance at a particular flow rate, (the design flow) gH can be calculated for design flow if values are selected for the other variables, or β_2 determined for a given value of gH.

Hydraulic efficiency η_h is selected from previous efficiency analyses for the type of machine. As has been explained, u_2 is proportional to N_2 and D_2, and N_2 is known at this stage. For example, a maximum value may have been calculated using an appropriate suction specific speed, k_{ss}, and the actual running speed matched to a prime mover. D_2 is the major geometric design variable and also appears in the expression for V_{m2}, associated with another design variable b_2, by the relation (allowing for outlet blockage due to the blades):

$$V_{m2} = \frac{Q}{b_2 [\pi D_2 - (zt/\sin\beta_2)]}$$

Normally, since H and Q are specified, the head equation can be solved for b_2 for a range of values of D_2 which are estimated using a modelling technique or head coefficient analysis. Acceptable proportions D_2/b_2 and b_2/B_2 can be approximately established from past experience of the influence of these ratios on efficiency and characteristic curve shapes over a range of specific speeds.

The remaining unknown, h_0, can be evaluated as described in Section 3.4.4.

The selection of a slip coefficient depends primarily on values of z and β_2, both of which are variable but within fairly narrow limits. Now that numerical computation is somewhat faster it is perfectly feasible to develop an interactive program, on say a desk-top computer, which will allow the designer to choose a range of values for each and calculate the corresponding b_2/D_2 variations.

That is not to say that there is not an element of empiricism in applying the data since those using the Busemann data generally make their own corrections based on long term experience of the performance of consistent designs. The head coefficient, h_0 (Figure 6.6) is really a slip coefficient predicted for radial flow, logarithmic spiral vaned impellers having a limited range of r_1/r_2 and solidity (or vane overlap). If h_0 is to be applied to non-radial impellers, eg, mixed flow, then a correction should be made for the lay-back of the impeller – see Figure 3.15. If the lay-back is measured on the mid-stream line, h_0 has been found to be reliable.

Corrected h_0 = Uncorrected $h_0 \times \operatorname{cosec} \phi$ \hfill (6.3)

Note that this expression tends to infinity when ϕ approaches zero, i.e., an axial flow machine, but note also that r_1 also approaches r_2 for this condition, thus the Busemann analysis is invalid.

This facility would assume greater importance when the vanes are laid out since the detail may require changes in the initial decisions, due to excessive blade loading for example.

Since the closed valve head is governed mainly by impeller diameter, D_2, and the duty specific energy rise is affected quite markedly by

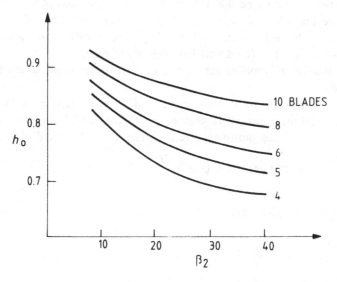

Figure 6.6. The variation of the Busemann coefficient h_0 with outlet angle β_2 and the number of blades.

z and β_2 (this can be tested by inserting ranges of values in the specific energy equation), the slope of the total head/flow rate curve can be controlled quite sensitively by firstly selecting values for D_2 and then varying z, β_2 and b_2. A very steep head/flow curve will involve a larger D_2 and hence a larger, less efficient pump giving a high test pressure rise for a given duty.

By assuming values for the design variables many 'suitable' designs can be generated to meet the required head and flow rate. It is then necessary to select the set of parameters giving the best compliance with all the design criteria. This involves further analysis, such as calculating vane loading and power curve manipulation to give a non-overloading characteristic, and empirical comparisons to assure acceptable efficiency and total head/flow rate characteristic shape.

6.5.2 Design procedure – area ratio method

As was discussed in Section 3.5, the basis of the area ratio method is a collection of normalised design data, in the form of head and flow coefficients plotted against area ratio, y, as seen in Figure 3.25.

The first step is to select an area ratio, y, appropriate to the required specific speed, k_s, and design/performance criteria. Generally $k_s \times y$ would fall in the range 0.8–1.2 for high efficiency but other criteria influence the selection, e.g.: $k_s y \ll 1$ for very steep head/flow rate curve, $k_s y < 1$ for stability, $k_s : 0.4$–1.2, $y < 0.75$ for non-overloading, $k_s : 0.8$–1.2, $k_s y > 1$ for a small diameter machine.

The number of impeller vanes, z, and vane outlet angle, β_2 are then selected.

From a plot of head coefficient and flow coefficient versus y (Figure 3.25) the appropriate head and flow coefficient can be selected for y, z, β_2 and the pump type (e.g. volute or multi-vaned diffuser),

Using the 'head' coefficient, $\psi = \dfrac{2gH}{u_2^2}$

$$u_2 = \frac{2gH}{\psi} \text{ for a given } gH \tag{6.4}$$

$$D_2 = \frac{60u_2}{\pi N} \text{ for a given } N \tag{6.5}$$

and the flow coefficient, $\quad QC = \dfrac{V_3}{U_3}$ \hfill (6.6)

Hence throat velocity, $V_3 = QC \times u_2$ and throat area, $A_3 = Q/V_3$.

Now $Y = \dfrac{\text{OABV}}{A_3} = \dfrac{0.95\,\pi D_2 b_2 \sin \beta_2}{A_3}$ \hfill (6.7)

therefore

$$0.95\,\pi D_2 b_2 \sin \beta_2 = y \times A_3$$

and

$$b_2 = \frac{y \times A_3}{0.95\,\pi D_2 \sin \beta_2}$$ \hfill (6.8)

This procedure establishes all the major outlet (impeller/collector) design parameters necessary to commence drawing a detailed design. The 0.95 factor is an allowance for vane blockage at outlet. In practice the actual blockage would be calculated, since

$$\text{blockage} = b_2 \times \frac{zt}{\sin \beta_2}$$ \hfill (6.9)

(Any intended outlet tip 'trimming' should be allowed for.)

Having produced a design, an estimate of the performance is required and here again the method assists in that the analysis naturally groups similar designs which can then be used to predict the new pump performance curves.

In essence area ratio 'says' that for any specific speed there is an infinite range of designs possible. The method identifies the design with geometric ratios, and therefore velocity ratios, which are most likely to give the desired characteristics.

6.5.3 Design method – design coefficients

The first authorative reference on the complete hydraulic design of a pump was produced by Stepanoff in his book *Centrifugal and Axial Flow Pumps*, published in 1976. The design procedures suggested are still valid as long as up-to-date, relevant, data are used and the book will continue to form a useful reference.

Stepanoff's design method is based on velocity and geometry analysis charts plotted against specific speed. The principal variables are head and flow coefficients: head coefficient $2gH/u_2^2$, flow coefficient V_3/u_2 or, alternatively, flow coefficient V_{m2}/u_2.

In Stepanoff's approach, design parameters (geometric and velocity ratios) of existing designs are plotted against specific speed, N_s. Each pump is identified by type, size, number of impeller vanes, etc, to enable consistent data to be used in subsequent new design parameter selection.

The charts do assume a consistency of design so that only pumps of similar type, construction, number of impeller vanes, etc, should be plotted on one chart. If Stepanoff's own data (1976) is used there is little design flexibility in that, superficially at least, only one acceptable design can be produced for each specific speed. If the whole text is studied in greater depth guidance is found on the effect of changing selected factors.

To produce a preliminary design layout, the vane inlet and outlet angles, blade number, meridional velocitiers at inlet and outlet, impeller outlet diameter and volute throat area are determined using a series of design charts, and empirical formulae.

6.5.4 A comparison of approaches to design

A revealing exercise was sponsored by the Institution of Mechanical Engineers in 1982. Pump design engineers were invited to predict, using their own methods, the hydraulic performance of an end suction volute type pump, a diffuser stage in a multi-stage pump, and a mixed flow pump of the bowl type. Five different approaches were involved: slip and loss analysis by Bunjes and Op de Woerd from the Stork company, the area ratio method by Thorne from Worthington Simpson, size shape and similarity formulae used by Hayward Tyler quoted by Richardson (1982), graphical and empirical methods introduced by Chiappe from W H Allen, and the NEL computer based methods outlined by Neal and by Stirling.

Since the present exercise is concerned with an end suction volute type pump it is interesting to compare the predictions made by the authors of that type of pump in 1982. These are shown in Table 6.1. The four non computer based predictions were all effectively in-house methods used by the companies from which the authors came, and they were comparing their results with the data from a pump designed and made by another company. Several of them commented that there appeared to be a mismatch between the impeller and the volute casing, and Thorne did not give an $NPSH_R$ prediction because he felt that the suction was over sized. The NEL methods gave the nearest prediction

Table 6.1. *A comparison of the predictions of several design engineers at the I Mech E Conference: 'Centrifugal pumps – Hydraulic Design', 1982.*

	Shut off Head m	Best Efficiency	Best η Flow l/s	Best η Head m	NPSH$_R$ at design Flow. m
Bunjes and Op de Woerd	56	78.5	56.9	44.8	5
Thorne	58.5	77	52.6	51	
Richardson	58.5	79	55	47	6
Chiappe	55	69.5	57.5	38.5	
Neal and Stirling (NEL)	58.5	80	60	48.5	8
Actual Test data	56	80	60	46	5

to that for the actual machine but gave the highest NPSH$_R$. All the approaches used were valid as they gave points in a narrow band. The pump designer is thus on firm ground if the approach used is based on well established data. In this approach the slip method has been used rather than the other methods, so the calculation will proceed.

6.6   Calculation of outlet diameter and width using the 'slip' method

$$\eta_0 = \eta_h \times \eta_{me} \times \eta_D \times \eta_v \qquad (6.10)$$

η_0 has been found to be 82.5% in Section 6.2, so inserting typical values of η_{me}, η_D and η_v allows η_h to be estimated. Since the product of the three efficiencies can be 95% for a pump of the size considered. and η_v (see Nixon and Cairney (1972)) could be about 1.2%,

$$\eta_{hyd} = \frac{0.825}{0.95} = 0.87$$

Thus, assuming 5 blades 3 mm thick $\beta_2 = 25°$, actual flow = 250/100 × 1.2 = 253 inserting in Equation (6.3) gives

$$V_{m2} = \frac{253}{3600} \frac{1}{0.023 \left[\pi\, 0.23 - 5 \times \dfrac{0.003}{\sin 25} \right]}$$

$$= 4.44\ \mathrm{ms}^{-1}$$

Since the estimated radius ratio = 1.7, $h_0 \simeq 0.75$ from Figure 6.6 –
Using Equation (3.7):

$$g\frac{55}{0.87} = u_2^2\left(0.75 - \frac{4.44}{u_2 \tan 25}\right)$$

$$\therefore u_2 = 35.71 \text{ ms}^{-1}$$

and

$$D_2 = \frac{35.71 \times 2 \times 30}{2900 \times \pi} = 0.235 \text{ m}$$

Thus, $D_2 = 0.235$ m, and $b_2 = 0.023$ m and the radius ratio is 1.88,
since the suction diameter was determined to be 0.125. Since the ratio
is close to that used, so the design will proceed, but for a detailed
design this calculation should be repeated using the new radius ratio.

6.7 A discussion of blade design considerations

Now that the major inlet and outlet parameters are established the
elevation profile, and blade passages can be drawn. It is usual to
determine the shape of one of the shrouds, either hub side or eye side
by reference to previous designs, although there are one or two
principles to observe:

(a) Generally a streamlined passage into and through the impeller is
     desirable from an efficiency point of view. Discontinuities in fluid
     velocity, either in magnitude or direction, are usually harmful to
     suction performance, stability and smooth running characteristics,
     as well as efficiency, particularly at flows away from design.
(b) The shorter the axial length of the impeller the better the mechanical
     design, as shaft span, bearing centres and pressure containment are
     all improved.
(c) The profile will vary with specific speed and quite often past experi-
     ence with successful designs is built on by modelling the shape of
     one of the shrouds. (This shape subsequently becomes a boundary
     stream-line).

Various techniques are available to ensure a smoothly changing
meridional velocity and profile through the impeller. The most common
manual design method is the inscribed circle approach illustrated in
Figure 3.14. The inlet and outlet circles are known and by calculation

or plotting a smoothly progressive meridional passage is obtained. Typical of the formulations used is:

$$b = \frac{b_1}{x\dfrac{(b_1 - b_2)}{ab_2} + 1} \tag{6.11}$$

where

- b = diameter of inscribed circle at required point
- x = distance from centre of b_1
- b_1 = diameter of inscribed circle at inlet
- b_2 = diameter of inscribed circle at outlet
- a = distance from centre of b_1 to centre of b_2.

For radial impellers x and a may be measured radially but for mixed flow, measurement along the mid-stream line is better. (This particular formulation is based on a rectangular hyperbola).

It is usual practice to give the shroud shapes so constructed a relatively simple geometrical definition, i.e., a series of radii with specified centre points, to aid pattern construction and impeller machining.

Considering now the inlet edges of the blades, once the inlet area is determined using the correct inlet velocity, the shape of the leading edge must be determined. As Figure 3.14 indicates, the leading edges move further into the eye as characteristic number increases. This leads to improvements in efficiency, suction performance and stability.

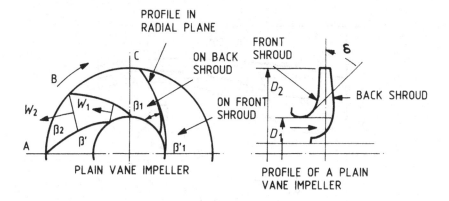

Figure 6.7.   A sketch of a simple centrifugal impeller illustrating the correction angle δ needed to allow for stream-surface curvature.

If a typical plain vane impeller, Figure 6.7 is examined, it will be observed that unless the liquid can change direction instantaneously at the vane inlet tip, the vane angles on and near the front shroud are incorrect. Figure 6.7 illustrates the point. Correction takes the form:

$$\tan \beta_1^1 = (\tan \beta_1) \cos \delta \tag{6.12}$$

Then if the space within the 'eye' is utilised by extending the vane towards the axis on the back shroud (see Figure 6.8) a change of radius of inlet tip also occurs. The two effects combine to determine the vane inlet angles on each stream-line. Some illustrations of the effects the position of the vane inlet tips have on pump performance are shown in Figures 6.8 and 6.9.

Once the position of the vane inlet edge profile is established, the vane profiles can be decided since the outlet geometry is determined. But blade thickness must be decided, since in most cases the process of manufacture is casting, which places limitations on section thickness and the way it varies. Best practice based on experience shows recommended minimum casting thicknesses for easy-to-cast materials. Mechanical strength and erosion resistance data for various impeller materials was given in Table 2.1.

Vane thickness creates blockage, in that the space taken up by vanes is not available to flow. Therefore the local velocities are increased, quite apart from the disturbances created by the vane inlet (and outlet) edges. Some authorities (e.g. Stepanoff) do not allow for inlet blockage where the vane inlet tips are faired, or streamlined. However, even

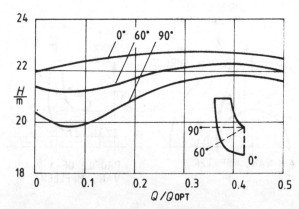

Figure 6.8. The effect of leading edge position on pump head rise.

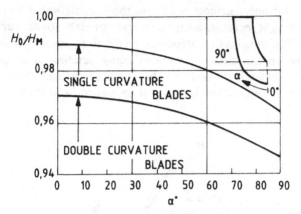

INLET EDGE POSITION AND STABILITY OF THE PUMP

H_0 - HEAD AT $Q = 0$

H_M - MAXIMUM HEAD OF THE PUMP CHARACTERISTIC

Figure 6.9. The effect of leading edge position and blade curvature design on pump head rise stability.

faired tips do have a finite tip thickness to interfere with inlet flow
so that some designers allow a half thickness blockage.

$$\text{Thus total inlet blockage} = \frac{z t b_1}{2 \sin \beta_1} \qquad (6.13)$$

and meridional velocities should be corrected accordingly.

Each position on the vane inlet edge can now be identified with a
local tangential velocity, u_1, and a local meridional velocity V_{m1} (V_{r1}
in the case of pure radial flow).

The local velocity triangle(s) can be constructed and the vane angle,
β_1, for shockless entry calculated from:

$$\beta_1 = \tan^{-1} \frac{V_{m1}}{u_1} \qquad (6.14)$$

It is common practice to add one or two degrees to this angle before
constructing the vane inlet shape to give a small incidence. This improves
suction performance and efficiency.

A similar calculation needs to be performed for each point on the
inlet edge intersected by the stream-surfaces selected to cover the flow
passage adequately. For low specific speed designs the hub and shroud

surfaces are the only surfaces used. As the specific speed increases (and passage height increases also) up to seven or eight stream-surfaces may be used. The construction often used is effectively one-dimensional, the area between the adjacent surfaces being determined to ensure equal flow increments through all the passages defined. The blade profiles are then determined for each stream-surface in turn.

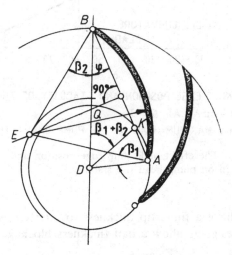

Figure 6.10. The single radius construction of blade profiles.

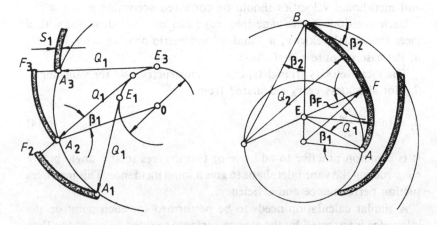

Figure 6.11. The two radius construction of blade profiles.

The circular arc may be used for very low specific speed designs (Figure 6.10). Its only virtue is simplicity, as the performance is not so good. The logarithmic spiral is the ideal flow shape, but the inlet and outlet angles are rarely the same. A construction method using 3 radii is shown in Figure 6.11.

A point-by-point method was introduced by Pfleiderer (1961) and also used by Church (1965). It is of use for higher specific speed impellers and is described in Figure 6.12. A method developed by Kaplan for water turbines known as the error triangle method is well described by Stepannof (1976) in Chapter 6 of his text.

Before proceeding with the numerical example, two further considerations need to be introduced. Both are related to the passages formed by the blade hub, and shroud profiles. The first is the angle of overlap, δ, shown in Figure 3.12. Early designs used a large overlap to control flow, giving a long passage. The optimum passage length is a matter of opinion, but remembering Kaplan's dictum that optimum geometry must give an acceptable balance between control and wetted area for good efficiency, an acceptable value for δ is $30°–45°$. The second consideration is to ensure that the passage area increases steadily from inlet to outlet (Figure 3.14) so that when the blade profiles are determined a check of area change must be performed, and if it is not smooth either hub and shroud profiles, or the blade shape, or both need to be adjusted to give an acceptable variation.

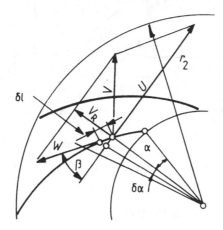

Figure 6.12. The point-by-point method approach to blade profile construction.

With the advent of the computer techniques the established methods used to assist the pattern maker, described in earlier texts, are being superseded by solid modelling and CNC programming.

6.8   The outline impeller design

6.8.1   Passage and blade shapes

The outer diameter, the outlet width, and the suction diameter are established and a decision made about the number of blades and outlet angle. The flow path geometry inlet angles and blade shapes must now be established.

For this trial design it was decided to allow a three-dimensional inlet profile, so assuming inlet normal velocity to be about the suction velocity, the passage height at 175 mm diameter, allowing for blockage

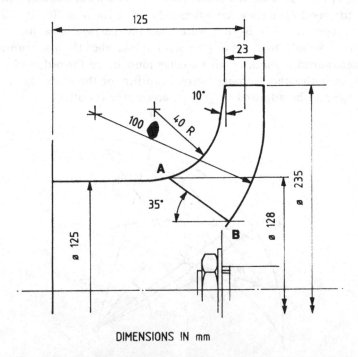

DIMENSIONS IN mm

Figure 6.13.   The calculated dimensions for the side elevation profile.

and a blade angle of about 20° was calculated to be about 28 mm. This allowed the shroud shape to be sketched, and the profile shown in Figure 6.13 emerged. The flow area was calculated for the inlet cone, to check that the normal velocity was close to the suction velocity. For the dimensions shown $V_{n\,inlet} = 5.3\,\text{ms}^{-1}$, close to the suction velocity.

For point A, $u = 19.46$, giving $\beta_{1A} = 14.5°$ suggesting a blade angle β'_{1A} of 12° if a small incidence is allowed. Similarly $\beta_{1B} = 23°$, giving $\beta'_{1B} = 25°$.

Allowing for the surface curvature as already discussed in Section 6.7, β'_{1A} and β'_{1B} in the vertical plane are respectively 5° and 16°.

Blade profiles may now be drawn as seen in the radial plane. Figure 6.15 shows 3 radius development for radial planes through points A and B. Using the method illustrated in Section 3.4, the passage area change should now be checked to ensure smooth relative velocity change from inlet to outlet. If this is acceptable the design can be detailed for patterns and machining.

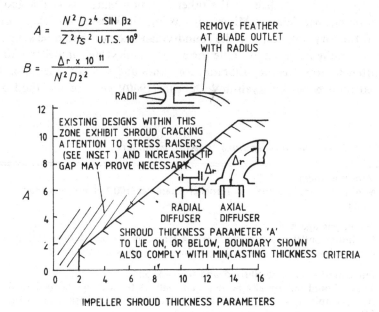

$$A = \frac{N^2 D_2^4 \, \text{SIN} \, \beta_2}{Z^2 t_s^2 \, \text{U.T.S.} \, 10^9}$$

$$B = \frac{\Delta r \times 10^{11}}{N^2 D_2^2}$$

REMOVE FEATHER
AT BLADE OUTLET
WITH RADIUS

RADII

12

EXISTING DESIGNS WITHIN THIS
ZONE EXHIBIT SHROUD CRACKING
10 ATTENTION TO STRESS RAISERS
(SEE INSET) AND INCREASING TIP
8 GAP MAY PROVE NECESSARY

Δr

A 6

4

2

0

RADIAL   AXIAL
DIFFUSER   DIFFUSER

SHROUD THICKNESS PARAMETER 'A'
TO LIE ON, OR BELOW, BOUNDARY SHOWN
ALSO COMPLY WITH MIN. CASTING THICKNESS CRITERIA

0   2   4   6   8   10   12   14   16

IMPELLER SHROUD THICKNESS PARAMETERS

Figure 6.14.   A method of establishing shroud thickness (Ryall and Duncan (1980)). (With acknowledgement to the Institution of Mechanical Engineers).

The two-dimensional approach has been used here, and only for two stream-surfaces, which usually suffices for a small pump. For wider passages it is necessary to establish other stream-surfaces, and develop the profiles along all of them. The text books by Pfleiderer (1961) and Lazarkiewics and Troskolanski (1965) may be consulted for the empirical approaches to profile development and the NEL methods outlined by Stirling and Wilson (1983) followed.

Before moving on to consider volute design, vane and shroud thicknesses must be examined a little further. From the hydraulic standpoint the ideal vane thickness will be zero, but of course hydraulic loads must be taken and the vanes and shrouds must be thick enough to withstand them and also give structural integrity to the impeller. There are a number of computer based packages available that are deployed for high energy input machines so that a complete stress analysis may be made. For a simple machine of the type discussed in the worked example a check on disc stress is all that is needed for the back plate at the maximum diameter. One reason for this apparently simplistic approach is that normal casting thicknesses needed for good castability with cast iron or the other common materials give a good margin against failure. Minimum casting thicknesses are based on good foundry practice, and the foundryman's advice should be sought for new designs. As an example for a cast stainless steel impeller 5 mm is often regarded as the minimum for vanes and shrouds. Figure 6.14 based on a paper by Ryall and Duncan (1980), gives one method of

Table 6.2. *Design considerations to assist 'castability'.*

1. Vane thickness
   Minimum vane thickness = Impeller Diameter/100 but not less than 4 mm.

2. Outlet passage width
   Minimum outlet passage width = 3–4 (Impeller outside diameter)/100 but not less than 12 mm or larger than 10 times vane thickness.

3. Fillet radii – Vane to Shroud and to back plate
   These should not usually be less than half of the sections being joined (for example if the vane, shroud and back plate are all, say, 5 mm the fillet radius should be at least 2.5 mm).

4. Thickness of central dividing rib in a double suction impeller
   Minimum thickness should be at least 4 mm and the radius therefore at least 2 mm or half the thickness of the dividing Rib.

establishing shroud thickness, and Table 6.2 outlines some design rules
for castability.

The thickness of 3 mm allowed in the design example earlier ought
therefore to be examined and depending on the material increased to
4 or 5 mm, and if necessary stressed later.

6.9 The design of the volute

The important first step in the volute design is to determine the
throat area. Using the area ratio concept, Figure 3.25 indicates that
the value of y should be about 1.10. Since the impeller outlet area is
$5.2 \times 10^3 \, \text{m}^2$, the throat area should be $4.73 \times 10^{-3} \, \text{m}^2$. If the throat
section is a circle, diameter D_T

$$D_T = \sqrt{\frac{4}{\pi} \times 4.73 \times 10^{-3}}, \ = 0.078 \, \text{m}$$

This gives $V_{\text{throat}} = 14.53 \, \text{ms}^{-1}$.

Following the principle outlined in Section 3.5.1, volute areas will be
decided by proportion round the impeller and Figure 6.15 illustrates
possible volute sections. The cut-water radius was based on an empirical
relation $= 1.1 \times$ impeller radius,

so the value $= \dfrac{0.235}{2} \times 1.1 = 0.12925$, rounded to $0.13 \, \text{m}$.

In establishing the volute cross-section areas the constant velocity rule
discussed in Section 3.5.1 has been used, with the volute throat velocity
being used for all sections.

The radius to the inner side of the cut-water was established by
using the rule that it shall be equal to $1.1 \times$ impeller radius. This
rule is based on the need to provide an adequate clearance as a very
close radial clearance has been found to give rise to excessive noise,
and at low flows this can give rise to the recirculation damage noted
in Section 3.3.

A minimum casting thickness of 5 mm has been assumed (if the
fluid were not clean and had a high content of abrasives or was very
corrosive a value of probably 10 mm would be assumed as will be
discussed in Section 9.3). The nose radius of the cut-water will thus be
2.5 mm. There is some evidence that a blunt nose gives a better range
of flows into the diffuser section of the casing. The radius will not be
affected in this design.

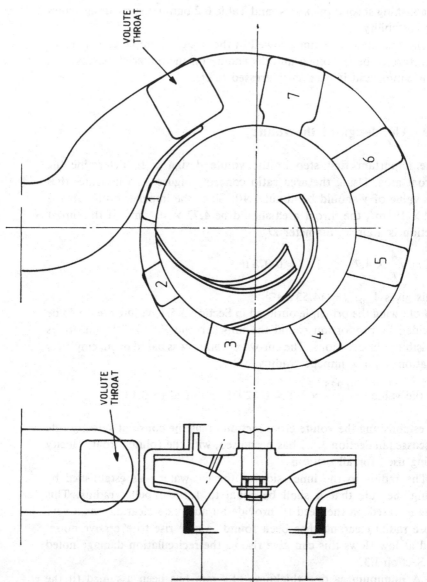

Figure 6.15. The plan view of the impeller and the volute casing with leading dimensions.

A volute wrap round of 360° has been assumed as can be seen in Figure 6.15, though there are some views that reducing this angle can give rather better diffuser design and allow for a better pressure recovery due to flow control being improved. As the Figure 6.15 indicates, the conventional diffuser shape has been provided to the discharge flange which is on the vertical centreline. The approach used for the cut-water shape is simple, and an alternative method is illustrated, by Stepannof (1976).

6.10 Radial and axial thrust calculations

6.10.1 Radial thrust

For the layout shown in Figure 6.15 Equation (3.10) will be used,

$$T_R = kp_2 D_2 b_2 \tag{3.10}$$

$D_2 = 0.235 \, \text{m}, \quad b_2 = 0.06 \, \text{m}.$

Following Stepanoff

$$K = 0.36 \left[1 - \left(\frac{Q}{Q_N} \right)^2 \right],$$

So

$K = 0.36$ at design flow Q_N.

$p_2 = 4.31 \times 10^5 \, \text{Nm}^{-2}$

Thus

$T_R = 0.36 \times 4.31 \times 10^5 \times 0.235 \times 0.06$

$T_R = 2188 \, \text{N}$

If a back wear ring is added as part of the axial thrust limitation package, as discussed below, b_2 will probably increase to 0.075 m, and $T_R = 2735 \, \text{N}$.

6.10.2 Axial thrust

Using the equations in Section 3.6.2, the axial thrust at design flow will now be determined.

$D_w = 0.14 \, \text{m}, \quad D_2 = 0.235, \quad D_s \simeq 0.04$

Thus $u_2 = 36.3\,\text{ms}^{-1}$
$\quad u_w = 21.6\,\text{ms}^{-1}$
$\quad u_s = 6.18\,\text{ms}^{-1}$
$\quad V_T = 14.53\,\text{ms}^{-1}$ from Section 6.9
$\quad p_s = 1.08 \times 10^5\,\text{Nm}^{-2}$
$\quad p_2 = 4.31 \times 10^5\,\text{Nm}^{-2}$ from Section 6.10.1

using equation (3.9)

$$T_{BP} = \frac{\pi}{4}[0.14^2 - 0.04^2][4.31 \times 10^5 - \tfrac{1}{8}]\left[36.3^2 - \left(\frac{21.6^2 + 6.18^2}{2}\right)\right]$$

$$= 6091\,\text{N}$$

T_s from Equation (3.12)

$$T_s = 1.08 \times 10 \times \frac{\pi}{4}(0.14)^2 = 1663\,\text{N}$$

$\therefore$ axial thrust $T_A = 4428\,\text{N}$

$$T_2 \;(\text{from Equation (3.14)}) = 10^2 \times \frac{250}{3600} \times 5.47$$

($V_0 = 5.47$ from Section 6.3) $= 380\,\text{N}$

Thrust on shaft (Equation (3.16))

$$T_3 = 10^5 - 1.08 \times 10^5 \times \frac{\pi}{4}(0.04)^2 = -10.06\,\text{N}\ (\text{that is in direction of } T_A)$$

$\therefore$ net axial thrust $= 4058\,\text{N}$ towards the suction pipe.

This can be reduced by providing a balance chamber, or by other means as discussed in Section 3.6.2. Practice differs when providing a rear wear ring with a balance chamber as illustrated in Figure 3.30; some designers ensure both wear rings are the same size, others provide a rear wear ring of different diameter to ensure a positive thrust at all times and thus reducing the risks of 'shuttling' and associated damage to rotating and static elements in the pump. Wear ring designs are discussed in the following section, and in Section 10.5.4.

6.11 Wear ring design

Figure 6.15 shows that in this design a balance chamber will be provided to give axial thrust balance. Also shown are two simple wear

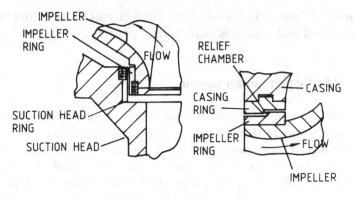

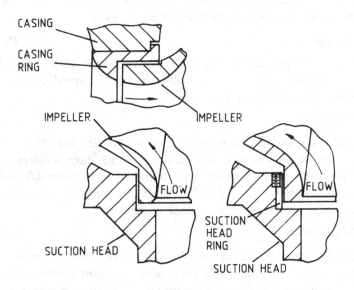

Figure 6.16. Types of wear ring used in centrifugal pumps.

rings that will be inserted with an interference fit and provided with a locking pin to prevent rotation.

There are several designs of wear ring used in pumps some of which are shown in Figure 6.16. All are designed to limit the flow from the zone at discharge pressure back to suction, so that the radial clearance acts as an annular orifice to control the flow by providing a pressure drop restriction. Clearly the radial clearance must be large enough to avoid contact between static and rotating members. This matter is discussed in Section 10.5.4 where the requirements of API 610 (1989) are discussed and compared with normal practice.

6.12   Multi-stage pump design

The worked example has concentrated on an end suction design where there was no doubt that a single-stage design was needed.

There is no precise rule for determining when a multi-stage design should be selected, or on how many stages are needed. A number of empirical rules exist, typical of these being those found in the Institution of Mechanical Engineers Manual edited by Davidson (1986). A simple first estimate of the number of stages is

$$Z_0 = H/7Q^{2/3} \tag{6.15}$$

this applies for pumps running at 50 revolutions per second. If sufficient NPSH is available, it is possible to raise the speed and reduce the number of stages using the rule

$$Z = (50/N)Z_0 \tag{6.16}$$

For horizontal pumps the manual suggests that limiting criteria are: number of stages $<620/N$, the maximum head/stage $<660\,\mathrm{m}$, and the dimensionless characteristic number will be in the range 0.03–0.08.

For small pumps where $Q \times H^{1/2} < 200$, or where the efficiency is not of prime importance the manual suggests that

$$2H/Q^2 \langle Z \rangle 8H/N^{4/3}Q^{2/3} \tag{6.17}$$

It must be remembered that as the number of stages increases the distance between bearings also rises, with attendant dynamic effects which become more and more important. Multi-stage pumps usually run at speeds above the first lateral critical speeds. The reader is referred to the manual mentioned above, and to API 610 (1989) for more information. Reference may also be made to the discussion on shaft design which is found in Chapter 8, and to the references cited.

6.13   Conclusion

The worked example will now terminate, as the process of detail design must take over on the route to the full design. The next steps must be to design the shaft system and the bearings using the thrust loads calculated. Decisions are needed on the correct seal system, and if double seals are required the flush system needs to be considered. Chapter 8 introduces the principles that must be followed for these elements, and also covers the selection of drives.

Pumps are used for difficult liquids which are corrosive or abrasive (or both), and in many cases solids are being pumped in suspension. The principles which underlie the correct design and selection of materials are discussed in some detail in Chapter 9.

Other points to be considered in the detail design are related to the effects of industrial codes and related matters which are covered in the last chapter as a summary of this approach to design.

7 □ The design of axial and mixed flow pumps

7.1 Introduction

Axial and mixed flow pumps will be treated as members of the same family, as they are high specific speed machines, even though the 'Francis' type of centrifugal machine has a mixed flow path. Both types exhibit the same characteristic behaviour, with a rise of specific energy towards shut valve which can be high in axials and in some mixed flow machines, and share a distinct tendency to unstable behaviour at part flow, and a power requirement which rises as flow reduces.

Energy is imparted to the fluid by blades rather than passages, and instability arises from flow breakdown over the blade profiles (giving rise to stall effects as described in Chapter 4). The interaction of fluid and pump components is complex, and performance is affected by blade profiles, surface finish, small variations in blade spacing and setting, and intake disturbances.

The principles underlying isolated blade profiles and blades in close proximity were described in Chapter 4, and simple but reasonably effective design techniques can be applied to the axial flow pump when the inlet flow is undisturbed. Quite efficient machines have been designed in the empirical way outlined in Section 7.3, but higher performance axial machines and all mixed flow machines require more sophisticated approaches, discussed in Section 7.4. Computer based methods were discussed in outline in Chapter 5.

7.2 An approach to design

The specific speed formula introduced in Section 1.7 indicates quite clearly that for a given duty, the faster the pump the higher the k_s. If k_s is higher the pump is smaller, lighter, etc. There is therefore an obvious major advantage in selecting a higher specific speed pump. The other reason for using high specific speed pumps is concerned with the head or pressure to be generated. The flow rate, Q, and $NPSH_A$ together determine the maximum, (i.e. economic) operating speed, as is indicated in Figure 7.1. If the head is low then the only way to reduce the specific speed would be to reduce the running speed to an uneconomic value, so that high flow rate results in an axial or mixed flow design rotating at conventional synchronous speeds, rather than the 'simple' centrifugal machine.

The fundamental difference between design for low specific speed and design for medium to high specific speed relates to the proportions of the total head or pressure rise generated by centrifugal action and blade lift. In impellers with few blades the application of the type of potential flow analysis carried out by Busemann breaks down and the actual 'slip' depends much more on the actual shape of individual vanes and the interaction between vanes (the cascade effect). Myles, (1965), suggests a re-arrangement of the Euler equation to illustrate the different components of total head rise for higher k_s pumps, with equation (1.3) becoming

$$gH_E = (U_2 V_{u2} - U_1 V_{u1})$$

$$gH_E = U_2(U_2 - U_1) + \omega R_2(W_{u1} - W_{u2}) - U_1 V_{u1} \qquad (7.1)$$

$$\underbrace{}_{\substack{\text{centrifugal} \\ \text{term}}} \quad \underbrace{\phantom{\omega R_2(W_{u1} - W_{u2})}}_{\substack{\text{vane} \\ \text{term}}} \quad \underbrace{\phantom{U_1 V_{u1}}}_{\substack{\text{inlet} \\ \text{whirl}}}$$

W_u = tangential component of relative velocity.

Unfortunately, although the correction to the centrifugal term is predictable using the relative eddy slip concept already discussed earlier, the vane cascade effects are not amenable to rigorous analysis. Many engineers have attempted to apply a universal model to the bladed cascade, e.g., Lieblein (1960), but without great success when applied to pump design. It has been found that the hub to tip diameter ratios, restrictions to blade number (due to stability and cavitation) and adjustments to vane shape from hub to tip (e.g. for mechanical strength) all tend to increase the empirical input into the design method.

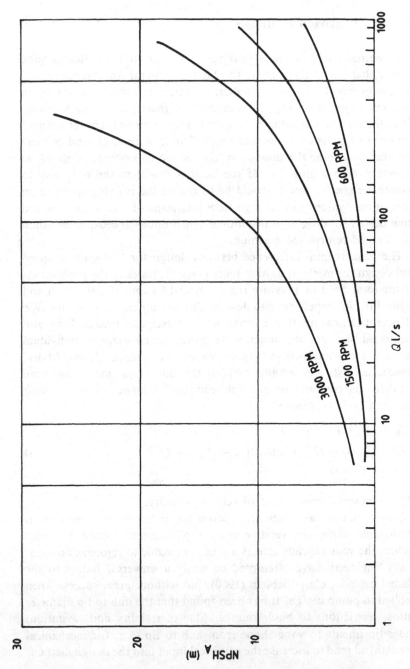

Figure 7.1. A design chart linking NPSH, rotational speed and flow rate.

The methods employed for mixed and axial flow pumps therefore tend to fall into two categories – those heavily reliant on empirical data and those based on highly advanced fluid flow analysis, which demand very powerful mathematical and computing capabilities. Both design approaches begin with the Euler equation, devise a suitable flow path, and then diverge in determining the blade profiles to be used. The well-established empirical approaches to axial design are discussed in the next section, and the sophisticated computer based methods are outlined in Section 7.4.

7.3 Axial flow pump design, an empirical approach

For axial flow machines a procedure for adapting the lift coefficient data for various aerofoil sections, in cascade for high solidity impellers, is often used. The fundamentals of this approach were given in Section 4.2.2, so Equation 4.5 can be used as the basis of the method. As Pearsall (1978) illustrated, this relation can be developed and used in design as follows.

Axial pumps are conceptually perhaps the simplest to design as the flow can be represented by a two-dimensional flow on axisymmetric surfaces. Assuming a 'free vortex' means there is no head change radially, and the flow can be truly represented on axisymmetric surfaces. Constant meridional velocity is assumed across the annulus, and this leads to calculation of the rotational or swirl velocity from knowledge of the blade characteristics.

Equating torque to the change in angular momentum at radius R results in the relation

$$C_L = \frac{4\pi R}{z \cdot c} \frac{V_u}{V_a} \sin\beta_m \left(1 + \frac{C_D}{C_L} \cos\beta_m\right)^{-1} \tag{7.2}$$

Similarly by equating the thrust to the pressure force on the fluid the pressure rise is given by

$$\frac{\Delta p}{\rho} = \frac{(C_L \cos\beta_m - C_D \sin\beta_m)}{4\pi R} zcW_{mean}^2 \tag{7.3}$$

The Euler equation gives the theoretical head rise (assuming no inlet swirl) as

$$gH_E = UV_{u_2} \tag{7.4}$$

Pump specific energy rise,

$$gH = \eta_h U V_{u_2} \tag{7.5}$$

where hydraulic 'efficiency'

$$\eta_h = 1 - \frac{V_a}{V}\left[\sin^2\beta_m\left(\frac{C_L}{C_D}\right) + \cos\beta_m\right]^{-1} \tag{7.6}$$

Design is then by iteration. If head, flow and speed or NPSH are given, then losses can be estimated and the theoretical head calculated. The required V_u can then be determined and from it the velocity triangles and geometric shapes. Aerofoil data can then be used to find the section and setting to give V_u and head, C_L and C_D. The losses are then calculated and the efficiency, which completes the iteration.

Empirical data suggests the desirable number of blades, typically Figure 7.2 based on the text by Stepannof (1976) can be used. If there are few blades, and the space–chord ratio at the hub section remain above 1, isolated aerofoil data may be used, but if blades are closer, corrections are needed, and the worked example that follows uses an approach due to Hay *et al.* (1978) developed for fans and proposed for pumps.

Alternative empirical approaches will be found in texts like those by Stepannof, and Lazarkiewicz and Troskolanski (1965) among others, and they illustrate how experience has been distilled into design practice.

Consider now a simple axial flow pump design:

A vertical water pump is required to deliver 1 m³ per second with an energy rise of 500 J/kg. It is to draw from a sump with a minimum submergence of 6 m.

Consulting Figure 7.1, a rotational speed of 600 rpm is indicated, or allowing for slip, 580 rpm will be used. Thus $\omega = 60.74$ rad/s.

Thus $k_s = \dfrac{60.74\sqrt{1}}{(50)^{3/4}} = 3.23$

From Figure 1.9 $\eta_0 = 0.82$

Thus driver power needed $= \dfrac{1 \times 10^3 \times 30}{0.82} = 60.98\,\text{kW}$

If $\eta_{\text{MECH}} = 0.95\ \eta_{\text{HYD}} = 0.86$
From Figure 7.3 $K_A = 0.45$

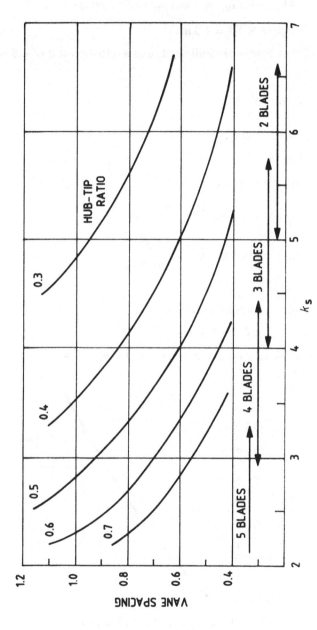

Figure 7.2. A design chart relating blade number, hub–tip ratio, and blade spacing (space–chord ratio). (Based on Stepannof (1976)).

Thus $V_A = 0.45\sqrt{2} \times 50 = 4.5\,\mathrm{ms^{-1}}$

using Figure 7.2, 4 blades are indicated, and a hub–tip ratio of 0.5 will be used.

$$\frac{1}{4.5} = \frac{\pi}{4}[D_T^2 - (0.5\,D_T^2)]$$

$D_T = 0.614$ (round to 0.62 m)

Thus $D_H = 0.31\,\mathrm{m}$

$U_{TIP} = 18.82\,\mathrm{ms^{-1}}$, and $U_{HUB} = 9.41\,\mathrm{ms^{-1}}$

and $V_A = 1 / \dfrac{\pi}{4}[0.62^2 - 0.31^2] = 4.416\,\mathrm{ms^{-1}}$

The Euler specific energy rise $gH_E = \dfrac{50}{0.86}$

$\therefore gU_E = 58.14\,\mathrm{J\,kg^{-1}}$

At the tip section, assuming zero inlet whirl,

$58.14 = 18.82\,Vu_2 \therefore VU_2 = 3.09\,\mathrm{ms^{-1}}$

The resulting velocity triangles are shown in Figure 7.4

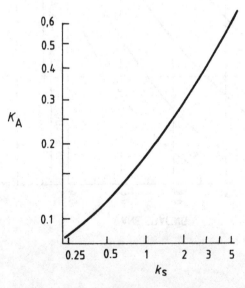

Figure 7.3. Axial velocity coefficient K_A variation with k_S.

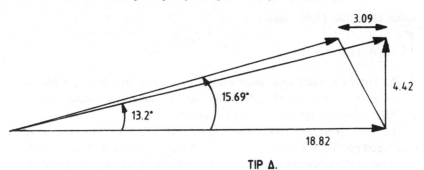

TIP Δ.

Figure 7.4.   Velocity triangles for the inlet and outlet edges for the tip section.

$\beta_1 = 13.2°$, $\beta_2 = 15.69°$ and $\beta_m = 14.45°$

using Equation (4.6) restated

$$\frac{gH}{U^2} = \frac{V_A}{2U} \cdot C_L \cdot \frac{c}{s} \operatorname{cosec} \beta m.$$

$$\left| C_L \cdot \frac{c}{s} \right|_{\text{TIP SECTION}} = 0.349$$

Similarly for the hub section, assuming free vortex distribution

$58.14 = 9.41 \, Vu_2; \, Vu_2 = 6.179$

Figure 7.5 results, and

$\beta_1 = 25.2°$, $\beta_2 = 53.83°$ and $\beta_m = 39:52°$

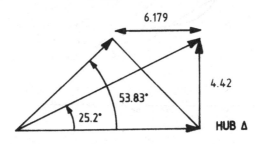

Figure 7.5.   Velocity triangles for the inlet and outlet edges of the hub section.

using Equation (4.6) again

$$\left| C_L \cdot \frac{c}{s} \right|_{\text{HUB SECTION}} = 1.78$$

The solution must now proceed to decide on the blade profile to be used, find the appropriate value for C_L and then, finding c/s, establish chord lengths and 'setting' angles.

The hub section shows a large change from β_1, to β_2, so a cambered profile will be required. The tip section shows a small change, so a low camber profile is indicated. A designer should use a consistent approach, but for illustrative purposes two approaches will be used here:

For the hub section an approach based on that proposed by Hay *et al.* (1978) will be used, and for the tip section a conventional low-speed wing profile utilised. Where $\beta_1 - \beta_2$ is large a large camber is needed,

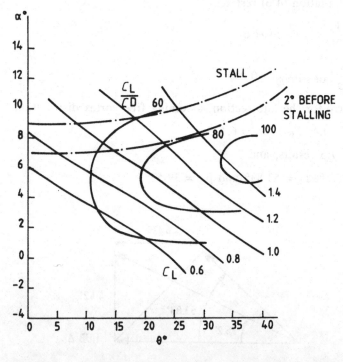

Figure 7.6.   Profile data for the C4 blade section replotted by Hay *et al.* (1978). Courtesy of the Institution of Mechanical Engineers.

and deviation as outlined in Chapter 4 must be calculated so that fluid angles can be correctly related to blade angles.

Hay *et al.* base their approach on the work of compressor pioneers like Howell (1945), and their methodology will be followed for the hub section.

As a start for the necessary iterative process of relating liquid angles with blade angles, it is assumed that $\beta_1 - \beta_2 = \theta$ the blade camber angle, so $\theta = 28.63°$.

In their paper Hay and his colleagues plot profile data in terms of θ and angle of incidence. For example Figure 7.6 shows the data for a common profile, the C4. Selecting data for 2° before stall (to give an operating margin) C_L/C_D is approximately 80, and C_L is 1.4, with $\alpha_i = 8°$.

Hay *et al.* define α_i by the relation $\alpha_i = (90 - \beta_1) - \gamma$. Since they refer fluid angles to the axial direction, α_i is thus equivalent to the angle of attack in isolated aerofoil data.

Thus the stagger angle $\gamma = (90 - 25.2) - 8 = 56.8°$.
Since $\varepsilon = \theta = 28.63$, referring to Figure 7.7 $\beta_1' = 20.97°$.

Using the Howell correlation for deviation (Figure 7.8),

$$\delta = (0.23\{2a/c\}^2 + \beta_2/500)\,\theta\sqrt{s/c}$$

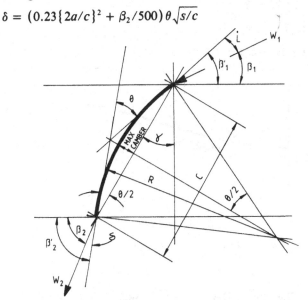

Figure 7.7. A comparison of liquid and blade angles.

since $\left| C_L \cdot \dfrac{c}{s} \right|_{HUB} = 1.78 \, c/s_{HUB} = 1.27,$

using a circular arc camber line ($a = c/2$ (Figure 7.7))

Since $\beta_2 = 53.83$

$\delta = (0.23 + 53.83/500)28.63\sqrt{1/1.27}$

$\therefore \delta = 8.58°$

Thus $\beta_2' = \beta_2 - 8.58 = 45.25$

A corrected value for $\theta = 45.25 - 20.97 = 24.28$
Returning to Figure 7.6 $\alpha_i = 7.5$
and $C_L = 1.22$, $C_L/C_D = 80$. $\frac{c}{s} = 1.46$
Thus γ is 56.3 and $\beta_1' = 23.22$
Repeating the calculation for δ

$\delta \left(0.23 + \dfrac{45.25}{500} \right) 24.28 \sqrt{\dfrac{1}{1.46}} = 6.44$

Thus $\beta_2' = \beta_2 - 6.44 = 47.39$

and $\theta = 47.39 - 23.22 = 24.17$

Thus from Figure 7.6 $\alpha_i = 7.5$

$C_L = 1.22$ $C_L/C_D = 80 \, c/s = 1.46$

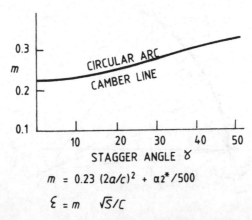

$m = 0.23 \, (2a/c)^2 + \alpha 2^*/500$

$\varepsilon = m \quad \sqrt{s}/c$

Figure 7.8. The Howell Deviation Correlation (Howell, 1945).
Courtesy of the Institution of Mechanical Engineers.

$\gamma = 56.3 \quad \beta_1' = 21.61$

$$\delta' = \left(0.23 + \frac{47.39}{500}\right) 24.17 \sqrt{\frac{1}{1.46}} = 6.5$$

$\beta_2' = 47.33 \quad \theta = 47.33 - 21.61 = 25.72°$

Iteration will cease as changes are small.
Thus the blade angles are:

$\beta_1' = 21.61°$
$\beta_2' = 47.33°$
$\gamma = 56.3°$
i = −3.59 well short of the C4 profile stall point
$c/s = 1.46,$

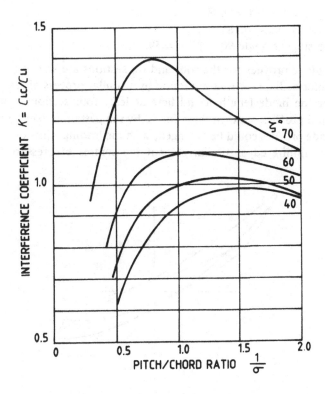

Figure 7.9. The correction for blade proximity proposed by Hay *et al.* (1978). Courtesy of the Institution of Mechanical Engineers.

$$\therefore c = 1.46 \times \frac{\pi \times 0.31}{4} = \underline{0.355\,\text{m}}$$

and the camber radius = 798 mm,

The interference factor in Figure 7.9 is approximately 1 so no correction to the above calculation is required, and the profile in Figure 7.10 results.

Turning now to the tip section and $\beta_2 - \beta_1 = 2.49°$. The C4 profile could be used on a very low camber but for illustrative purposes a conventional low speed flat bottom profile like that shown in Figure 4.6 will be used. Consulting Figure 4.6 C_L is 0.92 at $\alpha = 5.5°$, the angle of attack that corresponds to the maximum C_L/C_D ratio.

Thus $\left|\dfrac{c}{s}\right|_{\text{TIP}} = 0.349/0.92 = 0.38$

Spacing is $\pi\,0.62/4 = 0.487$
so the tip chord = 0.185 m
and the stagger angle $90 - 30.7 = 59.3°$.

The blade profiles for the root and tip sections are shown in Figure 7.10. Intermediate sections are found by a similar process of reasoning, and for the blade length found here at least four sections would be used, and profiles between them found by the pattern maker by ruling. The blade profiles could be arranged, with the leading edges on a radial line (or trailing edges similarly) or if the centres of pressure for all

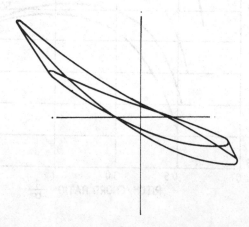

Figure 7.10. The proposed blade profiles superimposed.

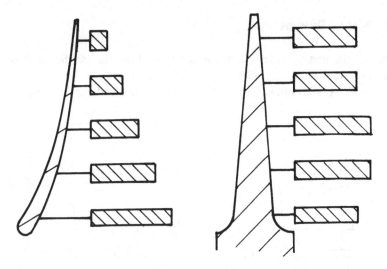

Figure 7.11. A schematic representation of the loads applied to a rotating axial flow blade.

sections are known, these can be laid along a radial line to minimise torque and stress.

Figure 7.11 suggests how hydraulic loading varies along a blade, and if several sections are used the stresses, particulary at the root section, can be determined to check that the root section will be adequate to withstand the hydraulic and mechanical loads implied by the duty for which the pump is being designed.

Many axial flow pumps are not given an outlet system designed to recover pressure energy, and in most cases are provided with simple vanes that remove the outlet swirl from the flow leaving the impeller, as shown for example in Figure 4.1. The outlet vanes are simple and have as a main function the provision of support for the bearing housing as shown in the figure. (In the case of the Bowl pump the passages are designed to give some diffusion in the way discussed already in Section 3.5.2.). The passage shapes are required to give sufficient mechanical support for the bearing housing and any seal that may be needed for the design configuration required.

Figure 7.12 illustrates a typical axial flow pump layout.

7.4   Mixed flow pumps

Many of the methods described for centrifugal pumps can be applied to mixed flow machines with appropriate adaptions.

Anderson (1980) indicates that the area ratio method can be used up

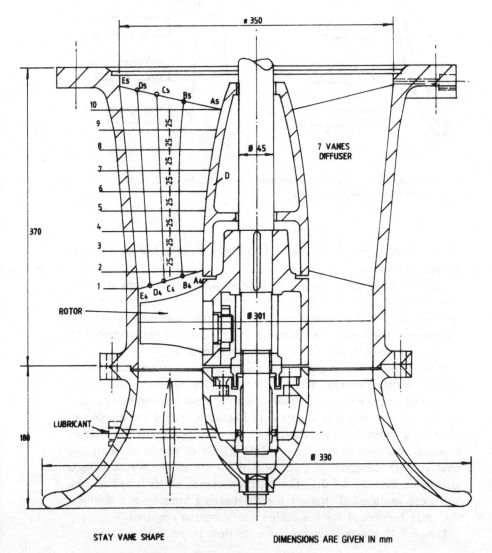

STAY VANE SHAPE   DIMENSIONS ARE GIVEN IN mm

Figure 7.12.   A typical axial flow pump.

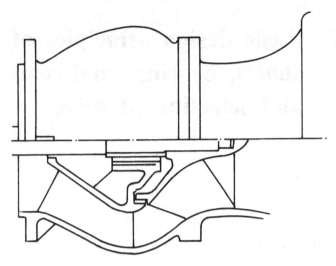

Figure 7.13.   A single stage 'bulb' pump.

to k_s values of about 4. The difficulty in using this method would be in establishing a reliable outlet area between vanes (OABV) since the very open construction (low solidity) of many impellers would make the definition and quantification a difficult geometrical problem.

Salisbury (1982) summarises the various approaches to the design of centrifugal pumps, with a comprehensive list of references. These apply equally to mixed flow pumps, with empirical factors for principal dimensions codified against specific speed or k_s. The reader is referred to the NEL reports by Myles, Stirling, Wilson and others which give a combined empirical and analytic design method which gives a good flow path and hydraulic design.

Figure 7.13 illustrates a typical bowl pump layout which will result from the approach.

8 □ Basic design principles of shafts, bearings and seals, and selection of drive

8.1 Introduction

It is not possible to cover anything in this book but the most basic considerations that underly the choice of shaft bearing seal and drive, so that the reader is referred to the literature and to pump handbooks that give more information. In this chapter the very basic principles are introduced, and the conventional terms used are defined.

Shaft design is first introduced, it being commented that this must involve the whole rotating system. Rolling element and plain bearings are then discussed, the basic seal designs are introduced and some design rules for good service are outlined. The chapter concludes with references to the selection of drive arrangement.

8.2 Shaft design

The shaft in a pump must sustain torsional effects, bending forces due to both the mechanical parts and hydraulic loads, and axial loads due to weight in the vertical plane and to hydraulic loads.

The empirical approach to shaft design is well documented in such texts as Stepannof (1976), Karassik *et al.* (1976) and the engineering handbooks. If the weight of the impeller system is known, and the axial and radial hydraulic loads determined, the shaft sizes can be determined and checked. Consider Figure 8.1, showing the rotating assembly for a horizontal centrifugal pump. Simple statics allow the determination of the reactions R_1 and R_2, and the resulting moment

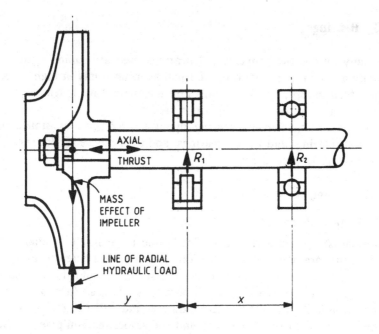

Figure 8.1. The rotating assembly of a horizontal centrifugal pump illustrating the load system applied to the bearings.

applied to the bearing system can be calculated. The empirical approach has led to satisfactory designs with large safety factors, but changes from packed glands (subsidiary 'bearings') to mechanical seals have meant that shafts designed in this way have failed. Increased knowledge of the fluctuating nature of the hydraulic loads, as described recently by Goulas and Truscott (1986), indicates that the fluctuating loads are very high particularly at low flow rates, and the shaft system is subjected to varying stresses and high vibration levels. This is recognised in such codes as API 610 (1989) which specifies vibration limits and that no 'critical' speed should occur in the operating speed range. Multi-stage pump design and vibration analysis has received much attention, and the reader is referred to the seminar paper by Lomakin (1958) and to papers by Brown (1975) and Kawata *et al.* (1988) as representative of experience and of a method of approach. The design of the shaft clearly involves a study of related components, wear ring clearances and rotordynamics, and computer programs using the accumulated experience are now widely used.

8.3 Bearings

In many pumps the conventional bearings used are rolling element, usually a combination of roller and ball bearings. Only in high speed boiler feed pumps, canned pumps and magnetic drive machines, for example, are hydrodynamic bearings used.

All the rolling element bearing manufacturers give assistance in selection, placing, and fits of bearings, and their advice service should be consulted in special cases.

8.3.1 Rolling element bearings

8.3.1.1 General considerations

Once the shaft size and bearing loads due to hydraulic, radial and axial thrust are determined, selection is effectively limited to makers catalogues, but bearing life and what designers should do to ensure it are not well covered in such publications. A large field of pump application is the Petro Chemical industry, and API 610 (1989) is the document used to specify pumps and the associated equipment. The code, among other things, specifies a nominal design fatigue life to be 3 years (25 000 hours) for continuous running at rated pump conditions, and at least 2 years at 'maximum' pump conditions of axial and radial loading.

In ideal conditions bearing life is determined by data on the fatigue of the rolling surfaces, which is susceptible to statistical analysis. Surface fatigue occurs as small pits or cracks which enlarge and multiply with time. These are very sensitive to load. Life varies inversely with $(load)^3$. Normal fatigue life (known as L_{10} life) is defined as the operating life which 90% of identical bearings running in the same conditions will reach or exceed. Figure 8.2 presented by Middlebrook (1991) illustrates data for several test bearings and shows that L_{10} life is well exceeded by many bearings. This life can be predicted by manufacturers to satisfy API 610, but only if all aspects of the fitting and use of the bearing are correct. Misalignment, damage in assembly, poor storage, lack of oil or grease, overlong lubrication intervals, ingress of water or debris, can cause accelerated damage, and other effects such as fretting, lack of internal clearance and cage breakage can and do affect life.

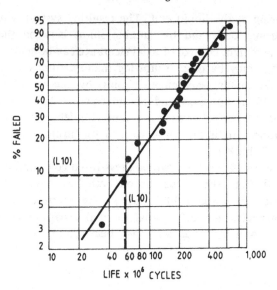

Figure 8.2. Fatigue life of a number of rolling element bearings. (After Middlebrook (1991)).

8.3.1.2 Bearing system design

The pump designer must minimise risk of failure by correct housing and shaft design, and also by the provision of adequate lubrication and isolation of the bearing system from external influences. He is, however, at the mercy of the user who maintains and services the equipment! Middlebrook (1991) itemises many factors which concern designer and user; here comments affecting basic design only will be made. The pump maker must provide the proper environment for the bearings and must ensure, after specifying an adequate bearing, that the housing, shaft and retention systems are correct.

Since radial and thrust loads must be sustained, and axial location of the shaft is important, a common simple system is to use two deep-groove ball bearings, locate one and allow the other axial freedom. This is adequate for very low axial thrusts, for example a 6132 ball bearing will take 850 lb (3781 N) radial, and 450 lb (2002 N) axial thrusts, for an L_{10} life at 3000 rpm, and the axial float is slight, much less than liquid end design usually needs.

For higher thrusts a design offered for process duties by a leading manufacturer is a pair of angular contact bearings at the drive end and a roller bearing at the liquid end. The resulting system will take the high fluctuating loads and by suitable shims between the angular contact bearings tight control of axial clearance can be achieved. Taper roller bearings may be used, following manufacturer's specifications as these offer similar capacities. 'Going by the book' requires a push fit of J6 in housings and an interference fit of K5 for ball bearings and M5 for roller bearings, but Middlebrook (1991) recommends tolerances to allow more control of fitting and refitting as tabulated in Table 8.1.

Bearings up to 70 mm bore can be press-fitted cold, and it is usually recommended that others must be oil bath heated to expand them enough to fit (up to about 90°C max.). The interference fit reduces internal clearances, and differential thermal expansions also do the same, so that API 610 specifies 'loose' bearings (Table 8.2).

Table 8.1. *Fits and tolerances for rolling element bearings.*

Bearing bore (mm)	Shaft tolerance K6 (m)	Housing bore tolerance J7 (m)
18–30	+15	+18
	+ 2	−12
30–50	+18	+22
	+ 2	+13
50–80	+21	+26
	+ 2	−14
80–100	+25	+26
	+ 3	−14

Source: Middlebrook (1991).

Table 8.2. *Normal and API 610 recommended tolerances compared.*

Clearance	150 BSI	Alternatively	
Least	2	C2 or 'one dot'	
Normal	(Normal)	'two dot'	
	3	C3 or 'three dot'	API 610
Greatest	4	C4 or 'four dot'	'loose'

After Middlebrook (1991).

8.3.1.3   Housing system design

Some makers prefer water cooled bearing housings when pumping a hot product. This gives rise to thermal relative expansion problems, and if there is any worry about heat soakage a cooling system for the oil is preferable.

Since the contact zone between ball and race is about 0.01 inch (0.025 mm) wide and the oil film only a few micro inches thick it does not take much roughness of surface to allow metal to metal contact, so filtration of the oil to keep out dirt is essential. Water contamination has been found to create problems, for example 400 ppm (0.04%) of water in oil will halve normal fatigue life and 100 ppm is about the lowest possible water content in a mineral oil. An oil with minimum water absorption capacity, like 'turbine' oil, and frequent oil changes are thus good practice, with the fitting of drain valves to allow dirt and water to be extracted.

Oil mist lubrication is favoured by some authorities in refineries, and of course a well maintained greasing system has its supporters.

Table 8.3 summarises the approach of one bearing specialist to a good bearing system.

8.3.2   Hydrodynamic bearings

8.3.2.1   Oil film bearings

Where the speed is beyond the normal limits for rolling element bearings, plain or hydrodynamic bearings are used. These are termed hydro-dynamic, with shaft and bearing separated by a film of oil in which

Table 8.3. *Rules for a good bearing system.*

1. Specify 40,000 hours L10 life.
2. Order 'good' bearings.
3. Specify C3 internal clearance (hot bearings over 50 mm bore).
4. Specify 40° contact angle paired angular contact bearings.
5. Fit them with 0.003 inch (0.076 mm) between inner races (to give good clearance).
6. Ensure precise fits in housing and on shaft.
7. Fit drain valves.
8. Go for coil or cartridge oil cooling.
9. Train fitters (and maintenance staff).
10. Provide for vibration monitoring.

Following Middlebrook (1991).

pressure is generated as sketched in Figure 8.3, and a load is thus sustained. The theory of such bearings is well established, with film thicknesses of the order of 0.0005–0.001 inch provided bearing and shaft are parallel. The important factors are the shaft finish (16 inch c.l.a.), alignment should be precise, and oil should be clean and debris free.

8.3.2.2 Water and product lubricated bearings

Many pumps used in river and bore hole pumps and for 'canned' boiler circulator pumps utilise plain bearings using rubber or other suitable materials for bearings. Fluted or grooved rubber is used where grit is present, and in boiler feed pumps fabric reinforced thermo-setting resins and other special material. Chemical pumps use graphite filled PTFE and other materials, depending on the chemical pumped. Table 8.4 taken from Middlebrook (1991) lists the suitability of various materials.

For plain bearings care in handling and fitting, high shaft finish, the correct fit to leave a small internal clearance, precise alignment, and the correct lubricant maintained in clean condition will pay dividends in prolonging the bearing life.

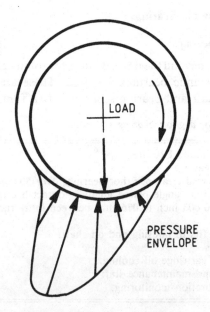

Figure 8.3. Pressure distribution in a hydrodynamic bearing.

Table 8.4. *Environmental suitability of polymer based bearing materials.*

		Acids	Alkalis	Solvents
Unfilled polymers	PTFE	Yes	Yes	Yes
	Polyacetal	No	Yes	Yes
	Polyamide (Nylon)	No	No	Yes
	HD polyethylene	No	Yes	Yes
	UHMW polyethylene	Yes	Yes	Yes
Filled polymers	Nylon + carbon fibre	No	No	Yes
	Polyester + carbon fibre	No	Yes	Yes
	UHMWPE + carbon fibre	Yes	Yes	Yes
	Nylon + MoS$_2$	No	No	Yes
	Nylon + Oil	No	No	Yes
	Polyacetal + oil	No	No	Yes
Filled PTFE	PTFE + glass	Yes	Yes	Yes
	PTFE + carbon/graphite	Yes	Yes	Yes
	PTFE + bronze	No	No	Yes
Reinforced thermosets		No	No	Yes

Source: Middlebrook (1991).

8.4 Seals

Three categories of seal will be discussed; mechanical, packed glands, and lip seals. Other seals are not suitable for sealing between rotating and static elements, and will not be discussed.

8.4.1 Single mechanical seals

The basic mechanical seal was designed to overcome the problems associated with soft packed stuffing boxes. Although designs of seal vary between manufacturers they all contain the same basic elements as shown in Figure 8.4.

To operate successfully with the minimum amount of face wear, the mechanical seal relies on the formation of a stable fluid film between the faces. In the majority of cases a liquid film of the product sealed exists between the faces.

The simple seal shown in Figure 8.4 is limited to fairly low pressure differentials and so for higher differentials a balanced seal is used, and Figure 8.5 shows how this balance is achieved.

Due to the rubbing action of the faces heat is generated, and in order for a liquid film to remain stable it is vital that this heat is conducted away from the faces. The heat generated from the faces is conducted through the seal assembly to the bulk product in the sealing chamber. In order to dissipate this heat some means of cooling has to be applied. The most common means is via a tapping from the discharge branch of the pump which provides a circulation flow of the product to the seal faces which is relatively cool. This circulation flow is then returned to the suction, either via a separate tapping in the suction line, or back via balance holes in the impeller. An alternative sometimes used is to cool the seal chamber via a cooling jacket. The circulation arrangement has the advantage that it prevents a build-up of sediments round the seal face area.

For any seal for a given sealed pressure and product, there is a maximum product temperature in the seal chamber above which the heat generated at the faces is not dissipated effectively and the liquid film between the faces becomes unstable. If the product temperature is greater than this maximum temperature then the normal circulation connection will not suffice to maintain film stability. It then becomes necessary to cool the product in the seal chamber to a temperature at which the seal is stable. To do this a cooler, either air or water, is inserted in the circulation line. When a large pressure differential

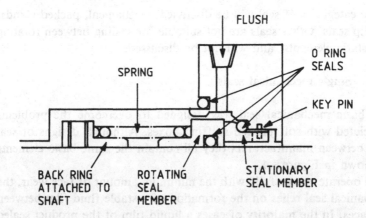

Figure 8.4.   A basic mechanical seal.

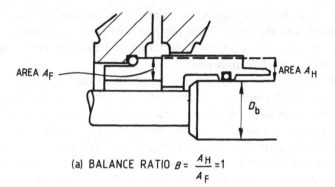

(a) BALANCE RATIO $B = \dfrac{A_H}{A_F} = 1$

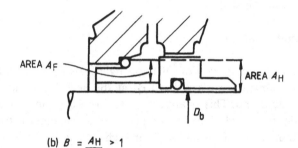

(b) $B = \dfrac{A_H}{A_F} > 1$

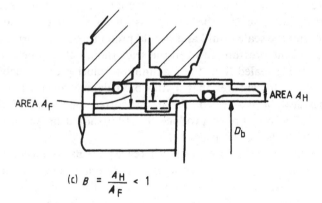

(c) $B = \dfrac{A_H}{A_F} < 1$

Figure 8.5. Force balance in a conventional mechanical seal.

exists between the two connected points in the circulation line there is a danger that the large flows produced will erode the seal rings. To prevent this a restriction such as a flow controller or orifice must be placed in the circulation line.

It is sometimes necessary to pump products containing abrasives in suspension. If these are allowed to get between the seal faces they can cause rapid wear and seal failure. One solution is to insert a cyclone separator in the circulation line; this supplies a flow of clean product to the seal, the dirty flow being returned to pump suction. A similar problem arises during plant commissioning when there is often a lot of dirt and pipe scale in the plant circuits. To prevent this entering between the seal faces a strainer should be put in the circulation line upstream from the seal.

8.4.2 Double seals

A double seal is a combination of two single seals fitted back-to-back in the seal chamber. One seal is fitted at the neck bush end and the other at the gland end. This arrangement (Figure 8.6) permits an external source of liquid or sealant to be pumped into the chamber between the seals. A tandem system often used for high pressure is also shown in Figure 8.6.

There are two basic considerations affecting the use of double seals: the barrier liquid, or sealant, between the seals must be a stable liquid compatible with the sealed fluid and should preferably have lubricating properties, and the pressure between the seals should always be above the pressure being sealed (i.e., the maximum product pressure behind the inner seal). This ensures that a film of sealant is always present at the inner seal faces – essential for the correct operation of this type of seal. The pressure to produce this film should be about one bar greater than the sealed pressure.

Double seals are selected for duties where single seals are unsuitable. Examples are: high temperature duties where materials are not capable of withstanding elevated temperatures of up to 4000°C, coupled with the danger of 'coking' or solidification of the product beneath the seal; cold duty sealing where freezing of the atmospheric vapours can cause seal failures; gas duties due to the absence of the fluid film; vertical applications such as mixers, where the seal would have to work with a gas or vapour blanket beneath it; food and pharmaceutical duties where contamination cannot be tolerated. Sealants for such duties

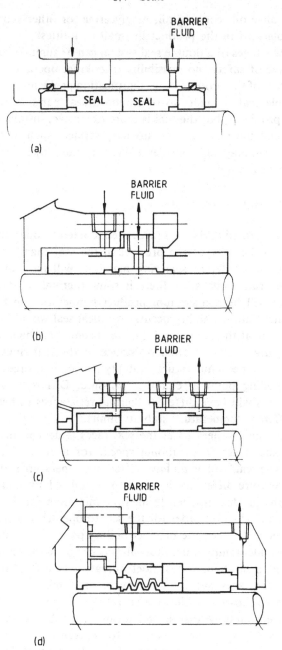

Figure 8.6.   Double mechanical seal layouts.

could be olive oil, cooking oil, or glycerine, or other substances that can be tolerated in the product in small quantities.

The advantages of a double seal system can be summarised as: a very high degree of safety; no possibility of coking; operation is generally independent of the product, and solidification problems are eliminated. The double seal is made up of standard seal parts so there are no stocking problems, but the seal is more expensive, installation is more difficult and the provision of auxiliary services such as the external sealant system may add considerable cost to the project, as a complete extra chemical system is needed.

8.4.3   Mechanical seal selection

The heat generated at the seal faces is conducted through the seal rings and then to the bulk of the product in the seal chamber. It follows from this that different seal ring materials will have different heat conduction rates, and since there is some thermal resistance to heat flow there will be a maximum product temperature in the chamber above which film instability occurs. An ideal seal would be one which could operate at the boiling point of the product at chamber pressure. However, due to the thermal resistance of the seal rings this is not possible and the temperature stability is some degrees below the product boiling point at chamber pressure. Different face material combinations will have different thermal resistances to heat flow and hence different temperature stability limits.

Since the heat generated at the seal faces is dependent on product pressure and mean face rubbing speed, for a given shaft rotation, different size seals will again have different temperature stability limits. The temperature difference between the stability limit and the boiling point of the product is often termed the minimum for the special seal combination. Then, in order for the seal to operate in a stable region, the difference between the product boiling point at chamber pressure and the product temperature must always be above the minimum level.

Although a mechanical seal does operate with a fluid film between the faces, the faces wear and so have a finite life.

It has been shown that linear wear-rate can be related to the product of the face contact pressure and mean face rubbing speed (usually termed $P \times V$). Since the size of seal for a given shaft speed is implicit in V, this limit applies over a range of sizes for a particular seal design and material combination.

The mechanical seal manufacturers provide a very complete technical service and the I. Mech. E. Guide to seals (Summers-Smith, 1988) forms an excellent guide for the pump engineer.

8.4.4 Packed glands

The basic properties a packing should have for good sealing capabilities and trouble free operation are: compatibility with the working fluid at the working temperature; plasticity to conform to the shaft under the influence of the gland force; lubricant insoluble and immiscible with the sealed fluid; non-abrasive to minimise shaft wear; non-corrosive to avoid damage to shaft or housing; and wear resistance to minimise gland adjustment.

Over the past 100 years soft packings have been used for sealing applications and many of the materials used today can be classed as traditional. Newer materials such as PTFE filaments, graphite filaments, foils, glass fibres, alumina silica fibres, aramid fibres etc. are now coming into common usage.

Yarn packings are most common and incorporate various types of fibre constructed as follows: plaited, braided, twisted, and 'plastic'. Yarns in common use are: vegetable fibres, mineral fibres, animal-based materials and synthetic materials.

In plaited and braided packings reinforcing wires are used for operation at higher temperatures and pressures. Lead alloy can be used but usually only as an aid to high speed running and is restricted to about 280°C (500°F). For high temperatures, special woven mesh types of copper and aluminium packing are available. Typical maximum temperatures for wire reinforced packings are as follows: stainless steel –875°C (1500°F), copper – 800°C (1470°F) max, inconel – 1200°C (2190°F), monel – 595°C (1100°F), aluminium – 530°C (1000°F) and brass – 510°C (950°F).

Metal foils of lead, aluminium, white metal and copper are used to form packings. Combinations of packings having end rings of a fairly hard material are used in conjunction with softer rings to form a packing set. Many variations are possible dependent upon the sealing duty. Mixed packing sets are used on such duties as sealing gases, volatile solvents, abrasives, fluids etc.

8.4.4.1 Housing design

The soft packing is compressed into a stuffing box by the compressive action of the gland follower. This compressive effort can be obtained by a simple bolted flange (two bolts) or a screwed flange.

For good operating life, the shaft should have a surface finish 0.4 m c.l.a. or better and the stuffing box bore a finish of 1.6 m c.l.a. or better. A 15° lead in, about 6 mm (0.25 in) long, at the mouth of the gland assists the entry of the packing and obviates the risk of damaging the packing in the assembly operation (Figure 8.7).

Clearance between the follower and neck bush inside diameters and shaft should be such as to prevent extrusion of the packing under load. The clearance between follower and housing should be small enough to prevent contact of follower and shaft, for if the follower shaft clearance is too small, solids collecting in the gap may give serious

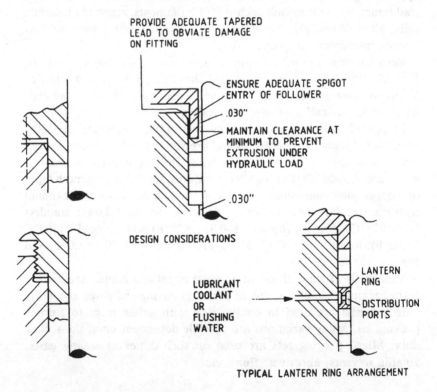

Figure 8.7.   Packed gland designs.

shaft abrasion. It is essential to provide adequate bearings for the shaft, as the soft packed gland should not be used as a bearing, though in many older designs with flexible shafts this tended to be the case.

Where extra lubrication, flushing, cooling or pressure balancing is required it is necessary to introduce a lantern ring (Figure 8.7). This is normally mid-way along a gland, and clean flush fluid is introduced into the gland via the lantern ring at a pressure slightly higher than the sealed pressure to reduce the possibility of abrasive particles coming into contact with the packing. Similarly the flush method is useful when sealing gases where toxic vapours present a health hazard. The flushing fluid can dilute the process fluid and this fact should be borne in mind when considering the use of a lantern ring arrangement.

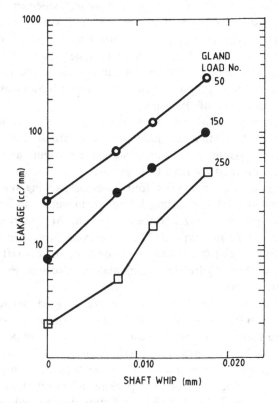

Figure 8.8. Leakage from a packed gland as a function of shaft whip.

Applications involving a medium which crystallises when cool can be troublesome, resulting in abrasion and even tearing of the packing immediately the shaft starts to rotate. This application can be dealt with by arranging a heating jacket around the gland so that the packing area is heated and the abrasive crystals liquified before start-up.

8.4.4.2 Performances

Leakage rates generally increase in proportion to shaft diameter and speed, and as the square of sealed pressure. Leakage is greater for intermittent operation of cycled temperatures or pressures.

For a pump operating with a shaft diameter of 25 mm, rotating at 3600 rpm with a fluid pressure in the stuffing box of 3.5 bars, and pumping clean cold water, the leakage would typically be 2 'drops/s' on first starting but falling to '1 drop' in 3 seconds when run-in. Better performance than this is possible with care and a good packing but some leakage is necessary to avoid seizure between gland and shaft. It should be borne in mind that leakage may be present although not visible. This is especially so with volatile liquids where the leakage escapes in the form of vapour which can present a hazardous environment in the proximity of the pump.

Leakage is very sensitive to shaft run-out or 'whip' (Figure 8.8). It must be remembered that the gland presents a hard or semi-hard bush to the shaft, is unable to accommodate shaft run-out, and becomes oval in time, followed by high leakage rates.

Typically a stuffing box with four rings of packing on a 50 mm diameter shaft at 1000 rpm sealing 3.5 bars cold water can be expected to exert a torque of 1.36–2.7 Nm when run-in. At a start-up before running in, 6.7–2.7 Nm is typical. The power consumption in the run-in state is typically about 0.2–0.4 hp. Increased speed, or shaft diameter would increase these figures in approximate proportion. Pressure has relatively little effect.

Wear rates are higher if the fluid contains solid particles or if the fluid tends to form crystalline solids on evaporation or polymerises at the seal to form solids. Shaft wear is usually more important than packing wear. Factors affecting shaft wear are: the fluid sealed (especially if abrasive or crystalline solids are in suspension); the shaft material; the packing material; the packing lubricant(s); the loading stress in the packing (related to the installation design) and shaft speed and stuffing box temperature.

The fluid sealed factor is probably the most important and can be alleviated by flushing clean fluid through the stuffing box, and using a harder shaft material (for example chromium plating, stellite, tungsten carbide or chrome oxide).

8.4.4.3 Operating limits

The operating limits on a packing are usually governed by temperature, rubbing speed, pressure and fluid.

The temperature range that packings will withstand is very wide. Suitable packings are available to work at extremely low temperatures, ($-200°C$), and extremely high temperatures ($+1200°C$). Some forms of PTFE packing will span the temperature range from $-200°C$ to $300°C$. The graphite filament foil type of packing spans the range of $0°C$ to $650°C$ (subject to the seal atmosphere).

Rubbing speeds of the order of $22\,ms^{-1}$ ($4000\,ft\,min^{-1}$) are handled satisfactorily by packings and a maximum of $30\,ms^{-1}$ ($6000\,ft\,min^{-1}$) has been quoted for boiler feed pump operations. The graphite filament packing due to its excellent thermal conductivity characteristics can run at very high speed [$30\,ms^{-1}$ ($6000\,ft\,min^{-1}$)], whereas the alumina silica packing (max working temperature $1200°C$) can only be run at a very low speed [some $0.2\,ms^{-1}$ ($4\,ft\,min^{-1}$)].

Pressures in excess of 69 bar (1000 psi) can be sealed with soft packings by paying attention to the dimensions and clearances of the stuffing box and gland follower at the required working pressure.

8.5 Selection of pump drives

Clearly the single speed electric motor driving a pump is the simplest system. Small pumps have the liquid end and motor as a unit, with the impeller on the motor shaft extension, and larger machines have direct or pulley drive systems.

Large multi-stage machines – boiler feed pumps for example – are now driven at higher speeds using steam turbines as drivers for economic reasons in power stations and for output regulation. Variable speed drives for smaller pumps can be electrical, mechanical or fluid systems, and such systems are now offered in many applications, both in water supply systems and in process industries. Bower (1981) presented in tabular form a comparison of available systems and Table 8.5 is based on his paper. His paper was written from the

Table 8.5. *Main characteristics of speed changing drives.*

Type of drive	Power range KW	Speeds		Drive efficiency %		Overall efficiency % including motor				Power factor		Main characteristics related to pump drives	
		Max rpm	Ratio	Maximum Speed	Half Speed	Max Speed		Half Speed		Max Speed	Half Speed	Advantages	Limitations
						4KW	150KW	4KW	150KW				
'V' Belts or flat belts	up to 750	5000 at Limited Power	8:1	'V' Belts / Flat Belts	85–90 / 90–95	70 / 75	80 / 85	40 / 45	65 / 70	0.9	0.3 with same motor	Low cost possibility if speed changes are infrequent.	Increased floor space. Not suitable for outside application. Reduced life if jack shaft not used.
Timing belts	up to 350	6000 at Limited Power	9:1	95+		80	87	50	75	0.9	0.3 with same motor	Similar to 'V' Belts but give reduced shaft loading and greater efficiency.	
Gear box	any	Any, but Standard Units usually step down		95+		80	87	50	75	0.9	0.3 with same motor	A robust, compact drive for infrequent changes or for matching drive speed to pump requirement. Any power available. Efficient.	Correct maintenance important.
Variable speed pulleys	up to 125	up to 4500 but at Limited Power	4:1 Std	85–90		70	NA	40	NA	0.9	0.3	Low cost. Housed-belt versions available.	Limited power range. Automatic control difficult. Limited belt life.
Fluid couplings	15 to 12 000	3500 Std	4:1	95	45–50	80	87	25	35	0.9	0.35	Reliability and life good. Available to very high powers at which they become more cost effective.	Low efficiency at reduced speed. Heat exchanger required for higher powers. Slip reduces pump max speed. Automatic control expensive. Expensive for low powers.

Mechanical variators	up to 75	4200 at 10 KW 2600 at 75 KW	up to 12:1	85–95	85–95	70	NA	40	NA	0.9	0.3	Speed increase possible.	Limited power range. Limited life. Careful lubrication and maintenance essential.
DC Motor & Thyristor voltage control	up to 2500	1500 Std, 3500 available higher-special	up to 30:1	see overall efficiency		80	90	45	65	0.9	0.3	Good range. Relatively good low speed efficiency. Automatic speed control easy.	TEFC Expensive. High speed expensive. Harmonics generation needs consideration. More maintenance than induction motors.
Voltage control of induction	up to 50	3000	4:1	see overall efficiency		80	NA	25	NA	0.9	0.3	Low cost. Can use squirrel cage induction motor. Automatic control easy.	Limited power range. Low efficiency at reduced speed. Only speed reduction possible. Requires at least 30% derating of motor.
Sohrage AC commutator motor	up to 500	2500	4:1	see overall efficiency		80	90	45	70	0.95	0.5	Good efficiency and power factor. Speed increase possible.	Automatic control expensive. Movable brush gear increases maintenance. TEFC expensive.
Stator-fed AC commutator motor	up to 2500	2500	up to 10:1	see overall efficiency		80	90	45	70	0.9	0.4	Improved life over the Schrage motor and increased supply voltages possible. Good efficiency.	Automatic control expensive. TEFC expensive.
Frequency control of induction motor	up to 75 Std	5000	10:1	see overall efficiency		75	87	40	65	0.95	0.95/ 0.3	Can use squirrel cage induction motors with minimal derating. Good power factor (with most designs). Speed increase possible.	High cost. Electronics more complex (and less reliable?) than other types.

Source: Bower (1981). Courtesy of The Institution of Mechanical Engineers.

view point of a pump designer and manufacturer, and he examined the advantages and extra costs critically. He concluded that a variable speed drive puts up original cost, but energy savings compared to valve control could give a payback time of 2 years in liquid transfer systems with mainly frictional losses.

Recently Shilston (1985) has discussed in some detail electrical drives, and Schwarz (1985) covers a wider range, with Chue and Lee (1983) at the same conference discussing an alternative twin engine drive. Alternative systems for providing variable flow are described by Waldron and Jackson (1983) and Lang and Rees (1981) and may be consulted to assist choice of system.

9 □ Pump design for difficult applications

9.1 Introduction

In addition to the pumping of clean, single phase liquids, pumps are used to transfer gas/liquid mixtures, suspensions of solids in liquids, sewage and multi-phase fluids. Most end suction designs are conventional in layout, as shown in Figure 9.1, with detail differences necessary to cope with erosion and high gas concentrations. This chapter discusses the solutions conventionally used to deal with aggressive chemicals, with hot and cold liquid pumping, with solids in suspensions, priming problems and high gas/liquid ratios.

9.2 Conventional process pumps

A conventional back-pull out pump is shown in Figure 9.1, which satisfies API 610. The liquid end consists of the impeller and volute, with shapes developed following the principles discussed in Chapter 6. The materials of the liquid end components used must be chosen to withstand the liquid being pumped as is discussed in later sections. The side and wear ring clearances used have been discussed in Section 6.11.

The seal and bearing systems available were discussed in Chapter 8. In this case a conventional mechanical seal provided with a flush connection, is used with a simple neck restrictor ring as a leakage limitation device on the seal failing. In all cases the seal type and system must be chosen to cope with the liquid properties, temperature and pressure. The bearing housing is conventional in layout, with a double row ball

161

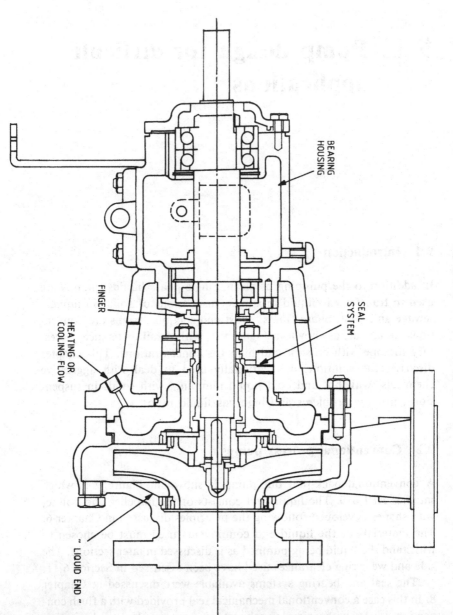

BEARING
HOUSING

FINGER

SEAL
SYSTEM

HEATING OR
COOLING FLOW

"LIQUID END"

Figure 9.1. A conventional back-pull out design satisfying API 610.

bearing to act as location and to take the hydraulic thrust loads, with an inboard roller bearing. An oil flinger is provided to ensure adequate lubrication.

It may be commented that operators need simple, robust, easily operated and maintained equipment, at low cost. Chemical pumps offered are required to pump liquids of relative density up to 2, viscosity up to 400 centistokes, and cope with temperatures from −90°C to 450°C with a two year bearing life specified as minimum.

9.2.1 Designing for corrosion/erosion

A selection of materials available for solid casings is listed in Tables 9.1, 9.2 and 9.3 together with some idea of compatibility with chemicals. With metals, an allowance is often used for corrosion and erosion of at least 3 mm on the normal 'water' thickness for conventional pressure levels, but it has to be remembered that wearing clearances can increase by a factor of 5 in use, with consequent reductions in volumetric efficiency, so drive sizing must allow for the increased power. With solid plastics, the designer can sometimes use conventional shapes used for metal castings, but often must settle for simple shapes like those used in glass pumps, as illustrated in Figure 9.2. Here the casing is in two parts clamped by bolts in tension to seal against the internal pressure. The efficiency of such designs is clearly less than good conventional centrifugal designs.

Table 9.1.

Cast Iron to BS 1452 grade 17 or 18/8 S.S.	Water, caustics (low temp) solvents
18/8 S.S.	Nitric acid, caustics, xylene, toulene and other solvents
18/10/Mo S.S.	Caustics, solvents, general chemicals, acetic acid, hydrogen peroxide
Hard rubber lining	HCL, barium chloride, sodium hypochlorate
Titanium	Sodium hypochlorite, chlorinated brines
Expoxy resin	Aluminium chlorohydrate, HCL, thorium sulphate

Table 9.2.

Liquid pumped	Conc. %	Temp. °C	Hastelloy	Polypropylene	Ryton	ETFE	PVDF	PTFE	99.5% Alumina silica
HCL	10	25	A	A	A	A	A	A	A
		100	A	–	–	A	A	A	A
	40	25	A	A	A	A	A	A	A
		100	B	–	–	A	A	A	A
H₂SO₄	20	25	A	A	A	A	A	A	A
		100	B	A	–	A	A	A	A
	50	25	A	A	A	A	A	A	A
		100	B	A	–	A	A	A	A
	98	25	A	A	A	A	A	A	A
		100	C	–	–	–	A	–	A
Nitric acid	20	25	A	A	A	A	A	A	A
		100	A	A	–	A	A	–	A
	98	25	C	A	C	A	A	A	A
		100	C	–	–	A	A	–	A
Caustic soda	10	25	A	A	A	A	A	A	A
		100	A	–	–	–	–	A	–
	70	25	–	A	–	A	–	A	–
		100	A	–	–	–	–	–	–
Benzene			B	D	A	A	A	C	A
Carbon tet. (wet)			A	D	A	A	A	A	A
Chromic acid	50		A	B	B	A	A	A	A
Phosphoric acid	70	25	B	C	A	A	A	A	A
		100	–	–	–	–	–	–	–
Ferric chloride			B	A	A	A	A	A	A
Hyd. peroxide	100		–	–	A	A	A	B	A

Key: A – Excellent   C – Fair
     B – Good   D – Unsuitable

Table 9.3.

Polypropylene
Produced using special techniques during the polymerisation process for pro-
pylene, results in a crystalline macrostructure. It has a good chemical resistance
to inorganic salts, mineral acids and bases; has useful mechanical properties
up to 100°C, good electrical properties, but resins unstable in oxidising conditions and under ultra-violet radiation. Easily processed using conventional thermoplastic techniques.

Ryton–polyphenylene sulphide
A crystalline polymer from benzene and sulphur; has good high temperature
properties (up to about 500°F), is hard at normal temperatures. Processed using
thermoplastic techniques.

PTFE
Sold as Teflon (Du Pont), Hostaflon (Hoechst) and Fluon (ICI). Resistant to
all known chemicals and solvents apart from molten alkali metals, gaseous,
flourine, and some halogenated compounds. Resins cannot be processed by
extrusion and moulding, but can be deposited in relatively thick coatings. In
rod form can be machined to close limits, relatively poor wear resistance can
be improved using fillers (carbon, metallic oxides, glass fibre etc), which also
give better rigidity and reduce cold flow under load.

ETFE
A copolymer of ethylene and polytetrafluorethylene (Tefzel Du Pont, Aflon
Cop, Asahi Glass). Fairly easily processed by conventional thermoplastic
techniques, can be heat-sealed and welded. Lower heat resistance than PTFE,
but higher tensile strength and toughness. Not attacked by most solvents up
to 200°C, good chemical stress-crack resistance.

PVDF
A fluoroplastic widely available Knar, (Pennwalt), etc. Good resistance to abrasion and cold flow, but chemically not as inert as PTFE or ETFE. No good
with ketones, esters, acetone, nitrobenzene, fuming sulphuric acid and other
similar solvents. Widely used for insulation of wire, in valves and piping. Easily
processed by usual methods, can be machined to close tolerances.

9.2.2   Hot/cold applications

Liquids such as tar products and sulphur are usually pumped hot, trace
heating in the circuits being commonly used; pumps must not be cooler
than the line so steam jackets are used as illustrated in Figure 9.3. The
temperature level means the seals need augmented flow, or double seals
with a flushing circuit should be fitted, and cooling is needed in the
oil system for the bearings as oil temperatures should not normally

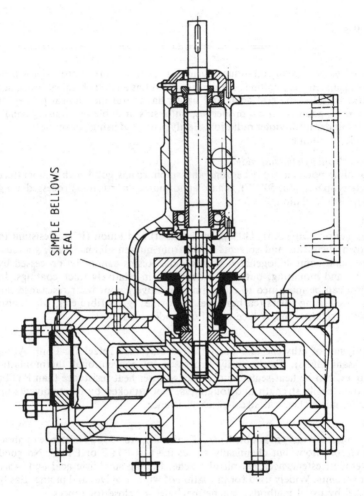

SIMPLE BELLOWS
SEAL

Figure 9.2. A simple glass pump.

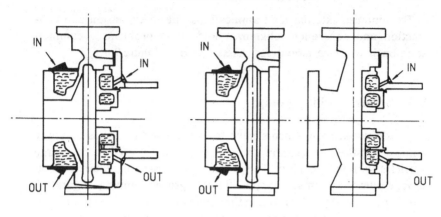

Figure 9.3. Schematic diagrams showing steam Jackets (the jackets could be hot water or cold water type depending on the pumped fluid).

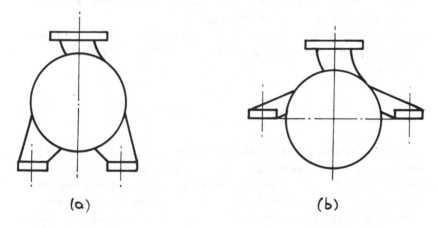

(a)   (b)

Figure 9.4. (a) Conventional and (b) centre mounted systems of pump support. Type (b) is used to minimise stress due to expansion problems.

exceed 80°C. If very cold liquids (such as cryogenic applications) are involved the jackets may be refrigerated to avoid 'hot spots'. In hot or cold applications running and stationary clearances must be chosen to avoid contact so the designer must allow for the effect on material structure of the temperature levels (cast iron is brittle at low temperatures for example), and for relative expansion or contraction.

To minimise external load applications due to expansion and contraction and consequent alignment and stress problems a common solution is to place mount feet as sketched in Figure 9.4.

9.3 Solids handling pumps

Two additional requirements have to be satisfied by solids handling pumps; resistance to erosion and corrosion, and a need to pass large particles without clogging. This latter is an essential requirement in sewage pumps and in such applications as gravel pumping. Figure 9.5 illustrates the range of solids handling impellers available, and it is conventional to specify the size of sphere that can be passed.

Figure 9.6 shows a typical gravel pump. The design includes replaceable wearing parts, and permits easy access to the flow path using covers and a split casing. The figure shows a stuffing box seal, but alternative designs use mechanical seals.

Metal thicknesses tend to be at least twice those used for normal water pumping, and some makers offer nitrile rubber or other materials as linings.

9.4 Glandless pumps

Where environmental considerations require no-leakage, alternatives to conventional sealing systems are required. Pumps are thus required that have no seals, two main types will be discussed, the canned pump and the magnetic drive machine. In both cases the liquid end is of conventional design, and the motor or drive is unconventional, since no seals

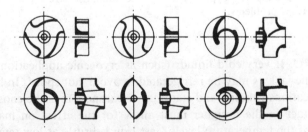

Figure 9.5. A range of solids handling impeller designs.

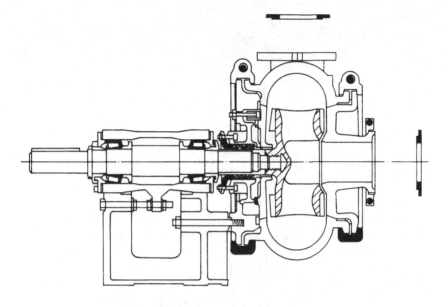

Figure 9.6. A typical gravel pump (based on the design offered by Simon Warman).

are used. The canned pump, Figure 9.7 may be vertical or horizontal, and uses product to lubricate the bearings, and if the product is not hot it provides cooling also. In many applications to which canned pumps are applied the product is volatile, so external cooling must be applied to the stator windings, and in many cases extra care is needed to avoid vapour locking in the motor. It must also be remembered that efficiences tend to be lower as illustrated in Figure 9.8. Figure 9.7 shows the motor placed above the liquid end, so that any gas liberated rises to the chamber above the thrust bearing, and so instrumentation is needed to ensure venting of gas before the pump runs.

A rather simpler design, but still seal-less, is the magnetic drive type, Figure 9.9 which uses a can, and the drive from conventional motor is through a magnetic ring system. It uses product for cooling and lubricating, and also has the canned pump vapour locking problem. The power penalty is less, as the use of rare earth magnets reduces slip to almost zero. The running radial clearance between the outer magnet and the can is of the order 0.5–1.00 mm, and similarly between the

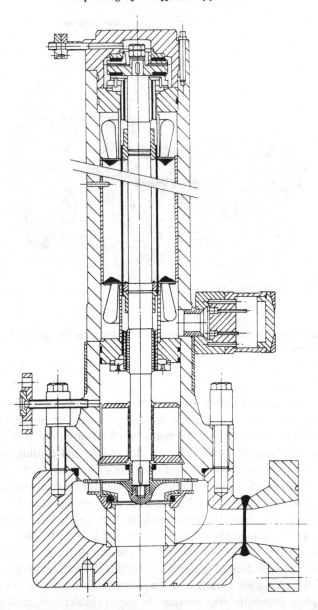

Figure 9.7. A canned pump.

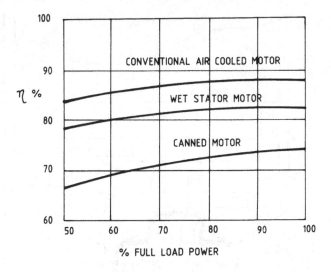

Figure 9.8. The efficiency reduction to be expected with wet stator and canned pump designs.

inner magnet and the can. Magnetic drive and canned motor pumps share the problem that the product pumped is the lubricant/cooling influence, however poor the lubricating properties of the liquid. The choice of bearing materials is thus crucial to the life of the pump, and they are carefully chosen. In many magnetic drive designs combinations of impregnated carbons are used where the liquid allows, and expensive ceramics where the liquid is aggressive.

In some cases where the pump is vertical a gas barrier is used to keep the liquid from rising into the drive, and as with other glandless designs the bearing system will be lubricated by the product.

9.5 Gas–liquid pumping

9.5.1 Priming systems

A centrifugal pump is primed when the liquid end flow passages are full of liquid, and the fluid is continuously flowing through the suction line and away along the delivery line. When a pump is first commissioned the suction line and the liquid end will be full of air, and before pumping can begin this air has to be expelled. In a drowned suction

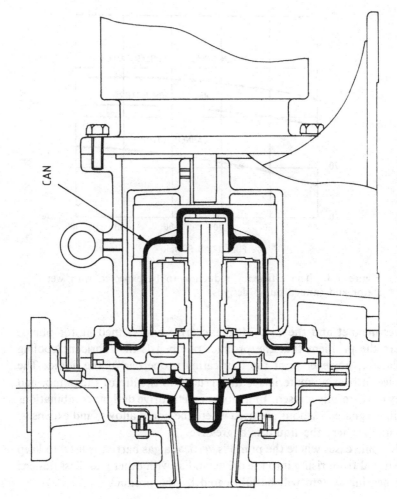

CAN

Figure 9.9. A magnetic drive type pump.

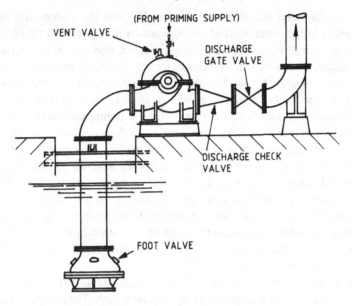

Figure 9.10. A typical suction lift system using a foot valve.

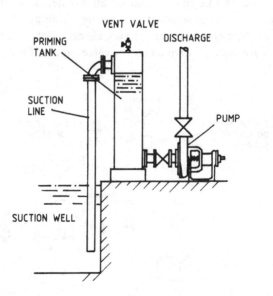

Figure 9.11. A simple priming system using a suction tank.

opening the suction valve will achieve this and the pump will begin delivering liquid when started. In the suction lift system, in tanker off-loading situations, and when pumping from pools or tanks open to atmosphere the pump needs assistance, as it cannot generate enough head rise to expel the air. Several solutions are available, the simplest being the foot valve at the open end of the suction line (Figure 9.10). This is closed in no-flow situations, and the pump and suction system is filled initially with liquid, the pump started, and when enough head is generated the foot valve lifts and flow begins. This works well with non-volatile liquids – the simple priming tank is an alternative. As Figure 9.11 shows the pump draws direct from the priming tank which has a volume about three times the volume of the suction pipe. When the pump starts it draws liquid from the tank, operating a vacuum in the tank free space; atmospheric pressure on the sump surface will then force liquid up the suction line. These tanks are usually used for small pumps, due to their size.

More complicated systems are those which use an external priming device, or depend on a special design of the pump. These latter devices are called self-priming pumps.

Two external devices are in common use, the ejector type (Figure 9.12) and the liquid ring vacuum pump (Figure 9.13). The ejector uses a primary source like the exhaust of an engine, or a compressed air supply, a steam line in some cases, or water from another pump. This then acts as the evacuating jet drawing air out of the pump until water fills the volute casing, when a control valve shuts off the primary flow.

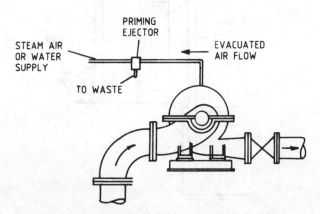

Figure 9.12.   The air ejector system.

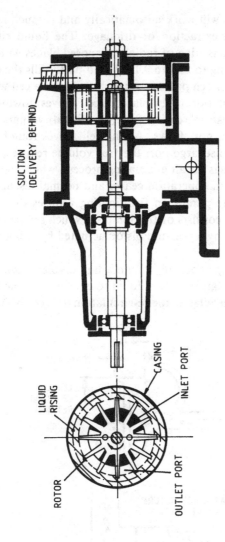

Figure 9.13. A schematic of a liquid ring pump used in centrifugal pump priming systems.

This sort of system will work automatically and is much used in contractor's site water extraction or drainage. The liquid ring vacuum pump has a rotor consisting of radially disposed blades within an elliptical casing. Referring to Figure 9.13, liquid partly fills the casing, and forms a liquid ring. Each pair of radial vanes forms a cell with the side walls, so that as rotation takes place, a cell moves downward across the inlet port with an increasing free (non-liquid) volume into which air from the suction port flows. As the cell moves round and moves upward across the discharge port the free volume reduces, forcing air out of the port. This is the air evacuating process, which continues until liquid flows, when the operation ceases and normal discharge occurs.

In the case of fire pumps on fire fighting vehicles the drive to the liquid ring machine consists of friction discs which are separated when the pump is primed by a pressure piston actuated by rising main pump delivery pressure.

Self-priming pumps normally retain a liquid volume when stopped, and provide re-circulation of this liquid to entrain air from the suction line. Two types are available: the re-circulation to suction, (Figure 9.14)

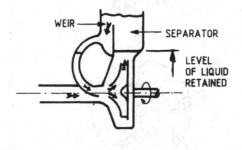

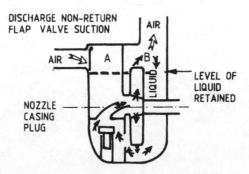

Figure 9.14. Re-circulation systems for self-priming pumps.

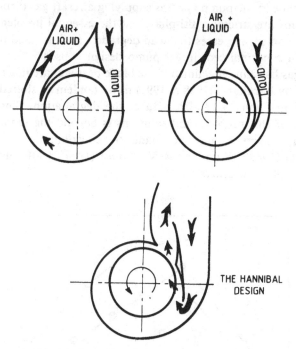

Figure 9.15. Three different peripheral re-circulation systems.

and peripheral re-circulation systems (Figure 9.15). There are a number of alternative layouts, but the basic principles are the same: the suction re-circulation system directs the retained liquid as it is discharged by the impeller into the suction zone, entraining air from the suction line, and the periphery re-circulation system directs the liquid back into the impeller preiphery where air is entrained and discharged. Both systems use a separator zone in the discharge casing where the liquid is decelerated, gives up air, and is re-circulated. There is obviously an efficiency penalty, but the pumps will run and re-prime as often as the duty requires. The handbook by Karassik *et al.* (1976) and other texts give more details, and recent contributions by Rachmann (1966 & 1967) and by Turton (1989) may be consulted.

9.5.2 Gas–liquid pumping

With off-shore oil exploration and wells there is now considerable interest in pumping liquids with a high gas content, as there has been for some time in the pump systems supplying aircraft gas turbines. Oil rig problems are strictly multi-phase but the essential problem is that as the gas content increases the head degrades, as illustrated in Figure 9.16, when a normal centrifugal pump design is used.

Better gas handling is claimed if the blade passages are much longer, and work by Furukawa (1988 & 1991) using tandem or slotted blades, illustrated in Figure 9.17 also reduces the degradation. Since a full discussion of this problem and solutions is beyond the scope of this discussion, reference may be made to contributions written by Kosmowski (1983), Murakami & Minemura (1974, 1980) Hughes and Gordon (1986) and others.

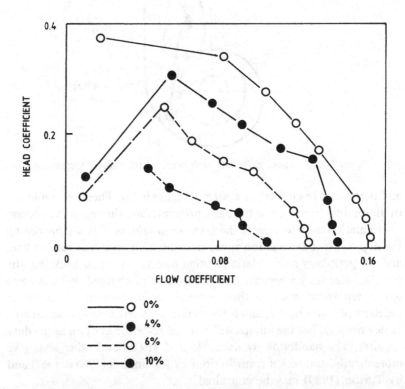

Figure 9.16. A comparison of the effect of air content rising from 0% to 10% by volume, on the head to flow characteristic of a centrifugal pump of conventional design.

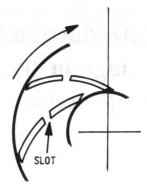

Figure 9.17. A sketch of a slotted impeller design intended to improve the gas handling capacity of a centrifugal pump.

9.6 Conclusion

This chapter has introduced the problems posed when a pump has to deliver liquids which are highly corrosive or are multi-phase, or has priming problems, and also ways of eliminating leakage are examined. The common solutions are outlined, and references given to allow further study.

10 □ An introduction to the next stage in the pump design process

10.1 Introduction

Chapters 1 to 7 have outlined the principles that underlie the successful design of the elements of the flow path in centrifugal, axial and mixed flow machines, and with the help of worked examples the logic to be followed has been explained. Chapter 8 has introduced the background to shaft design, seal and bearing system selection, and Chapter 9 has introduced the problems posed by difficult liquids, and the principles to be followed in choosing the correct materials and thickness of machine elements.

The first stage in the process of developing a pump is the establishing of the pump duty, which does not just mean the duty point but the fluid properties and other matters. Detail design is only complete when all the relevant standards and codes of practice are observed and satisfied. The discussion which follows covers these matters, and concludes with a brief discussion of test provisions and procedures.

10.2 Establishing the pump duty

An important factor in the process of producing the right pump design is the establishment of the pump duty. This demands a full interchange of information between the customer and the pump maker.

The pump designer requires to know, in addition to the rated flow rate, head and $NPSH_A$, the environment in which the machine is to be operated, the probable range of flow rates and heads that are to be

180

presented to the machine in the plant. Also needed are details of the fluid to be pumped, its properties at the rated duty and how they vary over the range of duties, what the level of contaminants is likely to be, and other data about the system that may be relevant such as the provision of filters and valves. Pump makers have produced their own questionnaires based on (sometimes) bitter experience, and the pump handbooks give examples of these.

API 610 (1989) is a good example of an industrial code of practice which includes a standard proforma to be completed by both customer and maker. It forms part of the formal contract between the two parties.

10.3   The design process

Once the duty head, flow rate and NPSH available have been established the design proceeds as follows:

The rotational speed must be determined. This is based on the preferred driver – electric motor or turbine. The specific speed (or characteristics number) is calculated and based on data in Section 3. The suction specific speed is checked, and the rotational speed chosen to fit the NPSH available. Tabular and graphical methods as used in Section 7, are also available.

The number of stages now needs to be found as explained in Section 6.12. Once this is determined hydraulic design proceeds as explained in Chapters 3 and 6 for centrifugal pumps, and in Chapters 4 and 7 for axial flow pumps. Once the flow path geometry is determined detail design proceeds, so that working drawings and specifications can be produced, costings done, and the pump manufactured.

To produce the detail drawings the casing and impeller proportions have to be determined, the seal system chosen, the bearings and shaft design selected. These, together with running clearances, and materials to be used are all dependent on the duty and fluid, and also the appropriate codes and standards. The influence of these will be discussed in the following sections.

10.4   The influence of dimensional standards

For European manufacturers supplying end suction pumps that deliver up to 16 bar rated pressure rise ISO 2858-1975 gives complete external dimensions of the pump, including the drive shaft sizing, base plate bolt arrangements, suction and dischange flange sizes and position for a range of nominal duties. Each pump is characterised by three numbers; the first number is the inlet diameter, the second the outlet diameter, the third the nominal impeller diameter. Figure 10.1 shows the full dimensions for the 80:50:200 pump as determined by the standard. The philosophy is thus that any pump conforming to the standard will fit on the base plate, fit the piping layout, and accept the same coupling giving complete interchangeability. A similar standard used in the USA is ANSI/ASME B73 1M-1984 sponsored by the American Society of Mechanical Engineers. This also lists relevant dimensional standards and material (ASTM) specifications, with an appendix giving non-binding layouts for seals and associated pipework.

10.5   The implications of API 610

A code for the petroleum industry that originates from the USA (API 610 (1989)) is quoted worldwide and adhered to in all petroleum industries. It has as its objective the provision of pumps for the industry that are fit for purpose, and is regularly up-dated.

The code covers casing loads and design/stress procedures, flange connections, rotating elements, wear rings, seals, bearings and housings, shafts, dynamics and vibration limits, specifying where appropriate the method to be followed in establishing stress load and vibration levels. It makes recommendations for accessories, inspection, testing, vendors and customers data.

10.5.1   Casing design

The code requires that the casing be designed to withstand discharge pressure and must withstand loads from the piping system. The procedure for static pressure loading and the over-pressure that must be applied is specified. It lays down when split casings should be used, and the limits of pressure, bolting, and all detail casing fixing provisions.

10.5.2   Rotating assembly dynamics

The criteria to be satisfied by the rotating assembly for critical speed, lateral and torsional vibration limits are stated. This is extended to

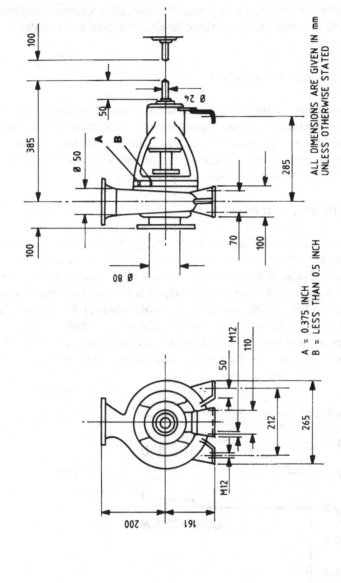

ALL DIMENSIONS ARE GIVEN IN mm
UNLESS OTHERWISE STATED

A = 0.375 INCH
B = LESS THAN 0.5 INCH

Figure 10.1. The leading dimensions of the ISO 2858–75 End suction pump denoted by the Code 80:50:200.

The next stage in pump design

include the whole driver–pump train. The code gives approved methods for calculating and checking these levels, and lays down balancing criteria.

10.5.3 Seal and bearing systems

Mechanical seals are preferred, minimum seal chamber dimensions are specified as are typical layouts discussed earlier in Section 8 of the code. Comprehensive specifications for the calculation of bearing and thrust loads are given, as are requirements for bearings and lubrication systems for both horizontal and vertical pumps.

10.5.4 Running clearances

Side clearances between impellers and casings are usually generous so that unless the thrust bearing fails pick up between rotating parts and casings is unlikely. Radial running clearances are closest in the wear rings so the code concentrates on specifying minimum clearances in the wear rings so that during warm up and cool down of hot pumps the opposing surfaces do not touch and thus risk pick up and seizure. The clearances are also there to avoid risk of pick up if the casing distorts. Renewable rings are usually fitted as discussed in Section 6, and the

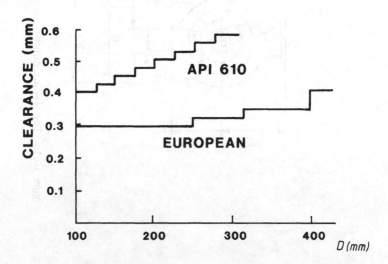

Figure 10.2. A comparison of conventional European wear ring clearances and those prescribed by API 610.

code specifies that materials having low galling tendency shall be used with clearances shown in Figure 10.2. Also shown in Figure 10.2 are the common running clearances used in European designs that are not to API 610, indicating that since they are much smaller the adherence to API 610 imposes a penalty of increased leakage flow (or lower volumetric efficiency).

10.5.5 Test provisions

Since the pump must be shown to give duty the code outlines the necessary work that must be done to establish its performance. Sections 4.3.3 and 4.3.4 in API 610 1989 detail the provisions.

The pump must be tested using water at 66°C maximum, using the contract seals and bearings, and complete assemblies. The hydraulic performance has to be found by using at least five data points including shut off, the rated flow, minimum continuous stable flow, midway between this flow and the rated flow and at 110% of the rated flow, with the test speed being maintained at a value within 3% of the rated speed stated on the contract. All the data obtained shall be logged and in most cases the tests inspected. Tolerances on head at shut off are ±10% for head up to 152 m; ±8% for heads from 153 to 305 m, and ±5% for heads over 305 m. The corresponding tolerances at the rated point are +5% to −2%: +3% to −2% and ±2%. Rated power has a tolerance of +4% at all points, and $NPSH_R$ is given no tolerance at all. $NPSH_R$ is to be found at minimum continuous flow, rated flow, midway between minimum continuous flow and rated flow, and at 110% of rated flow.

10.5.6 Other provisions

The code also covers drives, couplings, drive guards, baseplates, and all ancillary equipment that is part of the function of the pump, such as instrumentation. Appendices lay down procedures for calculating damped unbalanced response analysis and residual unbalance. Recommended data sheets and inspection check lists are also provided.

10.6 Concluding remarks

The extended discussion of API 610 (1989) illustrates clearly the amount of detailed work that has to be done to provide a pump that will meet the onerous safety demands of the petrochemical industry. Because it is so exhaustive there has been a tendency to use the code as a basis for contract in other industries that do not pose such hazards. This often leads to over specification and as a result has the effect of increasing cost.

Whether the pump is intended for the water supply industry, sewage systems, solids handling, pumping volatile chemicals, or heavy chemical duties, it must be 'fit for duty'. There is thus a responsibility placed on pump engineer and customer to ensure that a full and free exchange of data takes place. The pump engineer for his part must provide a machine that will provide reliable and economic operation.

The material in this book is an introduction to the process of providing the right design of machine, and the reader is advised to consult the sources of information quoted for detail design, stressing, material selection and other matters covered in outline in this treatment.

References

Abbot, I. M. and von Doenhoff, A. E. (1959). *Theory of Wing Sections*. Dover Press.

Ahmad, K., Lidgitt, P. J. and Dickson, H. M. K. (1986). A theoretical and experimental investigation of axial thrusts within a multi-stage centrifugal pump. *I. Mech. E. Conf. Radial Loads and Axial Thrusts on Centrifugal Pumps*, pp. 89–99.

American Hydraulic Institute (1983). *Hydraulic Institute Standards for Pumps* 14th Edition. Hydraulic Institute, Cleveland, USA.

Anderson, H. H. (1938). Mine pumps. *J. Mining Soc.* Durham.

Anderson, H. H. (1977). Statistical Records of pump and water turbine effectiveness. *I. Mech. E. Conf. Scaling for Performance Prediction in Rotodynamic Pumps*. September, pp. 1–6.

Anderson, H. H. (1980). *Centrifugal Pumps*. Trade and Technical Press.

Anderson, H. H. (1984). The area ratio system. *World Pumps*, p. 201.

API 610. (1989). *Centrifugal Pumps for Refinery Services*. 7th Edition. American Petroleum Institute, Washington, USA.

Balje, O. E. (1981). *Turbomachines: a Guide to Design, Selection and Theory*. Wiley.

Bower, J. (1981). The economics of operating centrifugal pumps with variable speed drive. *Proc. I. Mech. E.* Paper C 108/81.

Brown, R. P. (1975). Vibration phenomena in large centrifugal pumps. *I. Mech. E. Conf. Vibration and Noise in Pump Fan and Compressor Installations*. Paper C 101/75.

Bunjes, J. H. and Op de Woerd, J. C. H. (1982). Centrifugal pump performance prediction by slip and loss analysis. *I. Mech. E. Conf. Centrifugal Pumps – Hydraulic Design*, p. 17.

Busemann, A. (1928). The delivery head and radial centrifugal impellers with logarithmic spiral vanes (in German) *Z. Angew. Math. Mech.* **8** (5). p. 372. Translation into English by B. H. R. A. Cranfield.

187

188   *References*

Carter, A. D. S. (1961). Blade profiles for axial flow pumps, fans, and com-
pressors. *Proc. I. Mech. E.* **175**, 775-806.
Casey, M. V. and Roth, P. (1984). A streamline curvature through-flow
method for radial turbocompressors. *I. Mech. E. Conf. Computational
Methods in Turbomachinery.*
Chiappe, E. A. (1982). Pump performance prediction using graphical tech-
niques and empirical formulae. *I. Mech. E. Conf. Centrifugal Pumps -
Hydraulic Design*, p. 37.
Chue, Seck-hong and Lee, Yow-ching. (1983). Twin engine drive for pumps.
9th BPMA Tech. Conf. April, pp. 35-42.
Church, G. (1965). *Centrifugal Pumps and Blowers.* Wiley.
Davidson, J. (ed) (1986). *Process Pump Selection, a Systems Approach.* I.
Mech. E. Monograph.
Deeprose, W. M. and Merry, H. (1977). The effect of size and speed and
measured pump cavitation performance. *I. Mech. E. Conf. Scaling for
Performance Prediction in Applications.* Paper C 187/77.
Denny, D. F. (1954). *Leakage Flow through Centrifugal Pump Wear Rings.*
B.H.R.A. TN. 460.
Dixon, S. L. (1975). *Fluid Mechanics, Thermodynamics of Turbomachinery.*
2nd Edition. Pergamon Press.
Duncan, A. B. (1986). A review of the pump rotor equilibrium problem - some
case studies. *I. Mech. E. Conf. Radial Loads and Axial Thrusts on Cen-
trifugal Pumps.* pp. 39-52.
Dunham, J. (1965). *Non-Axisymmetric Flows in Axial Compressors.* I. Mech.
E.; Monograph in Mechanical Engineering (3).
Durrer, H. (1986). Cavitation erosion and fluid mechanics. *Sulzer Tech. Rev.*
3, 55-61.
Eck, B. (1973). *Fans.* Pergamon Press.
Frost, T. H. and Nilsen, E. (1991). Shut-off head of centrifugal pumps and
fans. *Proc. I. Mech. E.* **205**, 217-23.
Furukawa, A. (1988). Fundamental Studies on a tandem bladed impeller of gas
liquid two-phase centrifugal pump. *Memoirs of the Faculty of Engineering
Kyushu University*, **48**, 4, 231-40.
Furukawa, A. (1991). On an improvement in air/water two phase flow perfor-
mance of a centrifugal pump in the partial flow rate range of water. *69th
JSME Fall Annual Meeting*, Vol. B, paper number 1118, pp. 165-7 (in
Japanese).
Gongwer, C. A. (1941). A theory of cavitation flow in centrifugal pump
impellers. *Trans. A.S.M.E.* **63**, 29-40.
Goulas, A. and Truscott, G. F. (1986) Dynamic hydraulic loading on a
centrifugal pump impeller. *I. Mech. E. Conf. Radial Loads and Axial
Thrusts on Centrifugal Pumps.* 53-64.
Grist, E. (1986). The volumetric performance of cavitating centrifugal pumps.
Part 1, Theoretical analysis and method of prediction. Part 2, Predicted and

measured performance. *Proc. I. Mech. E.* **200** paper numbers 58 and 59.

Grist, E. (1988). Pressure pulsations in cavitating high-energy centrifugal pumps and adjacent pipe-work at very low or zero flow rate. *I. Mech. E. Conf. Part Load Pumping Operation, Control and Behaviour.* 143–51.

Hamrick, J. J., Ginsburg, A. and Osborn W. M. (1952). *Method of Analysis for Compressible Flow through Mixed Flow Impellers of Arbitrary Design.* NACA Report 1082.

Hay, N., Metcalfe, R. and Reizes, J. A. (1978). A simple method for the selection of axial fan blade profiles. *Proc. I. Mech. E.* **192**, 25, 269–75

Horlock, J. H. (1958). *Axial Flow Compressors.* Butterworth.

Howell, A. R. (1945) Fluid dynamics of axial flow compressors. *Proc. I. Mech. E.* **153**, 441–52.

Hughes, S. J. and Gordon, I. (1986). Multi-phase research for North Sea oil and gas production *J. Mar. Eng. Rev.*, pp. 4–9, July.

Hughes, S. J., Salisbury, A. G. and Turton, R. K. (1988). A review of CAE techniques for rotodynamic pumps. *I. Mech. E. Conf. Use of CAD/CAM for Fluid Machinery Design and Manufacture*, 9–19.

Karassik, I. J. (1981). *Centrifugal Pump Clinic.* Marcel Dekker. New York.

Karassik, I. J., Krutzsch, W. G., Fraser, W. H. and Messina, J. P. (1976). *Pump Handbook.* McGraw Hill.

Kawata, Y., Takata, T., Yasuda, O. and Takeuchi, T. (1988). Measurement of the transfer matrix of a prototype multi-stage centrifugal pump. *I. Mech. E. Conf. Part Load Pumping Operation, Control and Behaviour.* Paper 346.

Knapp, R. T., Dailly, J. W. and Hammitt, F. G. (1970). *Cavitation.* Engineering Societies Monograph. McGraw Hill.

Knapp, R. T. and Hollander, (1948). Laboratory investigations of the mechanism of cavitation *Trans. A.S.M.E.* **70**, 419–435.

Kosmowski, I. (1983). The design of centrifugal pumps for the delivery of liquids with high gas content. *8th BPMA Tech Conf.* Paper 13.

Kurokawa, J. and Toyokura, T. (1976). Axial thrust, disc friction torque, and leakage loss of radial flow machinery. *Int. Conf. on Design and Operation of Pumps and Turbines*, N.E.L. Scotland, September, paper 5.2.

Lakhwani, C. and Marsh, H. (1973). Rotating stall in an isolated rotor row and a single stage compressor. *I. Mech. E. Conf. Heat and Fluid Flow in Steam and Gas Turbine Plant.* April, 149–57.

Lang, V. and Rees, D. J. (1981). Economic advantages of using variable speed couplings in pump drives. *7th B.P.M.A. Conf.* March, 229–40.

Lazarkiewicz and Troskolanski. (1965). *Impeller Pumps.* Pergamon.

Leiblein, S. (1960). Diffusion Factor for Estimating Losses and Limiting Blade Loadings in Axial Flow Compressor Blade Elements. N.A.C.A. R.M. E53 D01.

Lewis, W. P. (1964). The design of centrifugal pump impellers for optimum cavitation performance. *Inst. Eng. Aust. Elec. Mech. Trans.* EM6 (2), 67–74.

Lomakin, A. A. (1958). Calculation of critical speed and securing of dynamic stability of hydraulic high pressure pumps with reference to forces arising in seal gaps. *Energmaschinostroenie*. Vol. 4, no. 1.

Lush, P. A. (1987a). Design for minimum cavitation. *Chartered Mechanical Engineer*. Sept. pp. 22-4.

Lush, P. A. (1987b). Materials for minimum cavitation. *Chartered Mechanical Engineer*. Oct. pp. 31-3.

Middlebrook, H. (1991). Bearing systems. Lecture given on 'Pumps in Service' short course School of Mechanical Engineering, Cranfield Institute of Technology.

Milne, A. J. (1986). A comparison of pressure distribution and radial loads on centrifugal pumps. *I. Mech. E. Conf. Radial Loads and Axial Thrusts on Centrifugal Pumps*, pp. 73-88.

Murakami, M. and Minemura, K. (1974). Effects of entrained air on the performance of centrifugal pumps (2nd report: Effects of the number of blades). *Bull.* J.S.M.E., **117**, 112, pp. 1286-95.

Murakami, M. and Minemura, K. (1980). Effects of entrained air on the performance of centrifugal pumps under cavitating conditions. *Bull.* J.S.M.E., **23**, 183, pp. 1435-42, Sept.

Myles, D. J. (1965). *A Design Method for Mixed Flow Pumps and Fans*. N.E.L. report 177, March.

Myles, D. J. (1969). An analysis of impeller and volute losses in centrifugal fans. *Proc. I. Mech. E.*, **184**, pp. 253-78.

Neal, A. N. (1980). *Through-flow Analysis of Pumps and Fans*. N.E.L. Report 669.

Neal, A. N. (1982). Theoretical methods for analysing the flow in pumps. *I. Mech. E. Conf. Centrifugal Pumps - Hydraulic Design*, p. 45.

Nece, R. E. and Dailly, J. M. (1960). Roughness effects on frictional resistance of enclosed rotating discs. *Trans. A.S.M.E., J. Basic Eng.* **82**, pp. 553-62.

NEL Course (1983). *Pump Design*. N.E.L. East Kilbride.

Neumann, B. (1991). *The Interaction between Geometry and Performance of a Centrifugal Pump*. Mechanical Engineering Publications.

Nixon, R. A. (1966). Examination of the problem of pump scale laws. *Conf. Pump Design Testing and Operation*. N.E.L., Paper D 2-1.

Nixon, R. A. and Cairney, W. D. (1972). *Scale Effects in Centrifugal Pumps for Thermal Power Stations*. N.E.L. Report 505.

Nixon, R. A. and Otway, F. O. T. (1972). The use of models in determining the performance of large circulating water pumps. *I. Mech. E. Conf. Site Testing of Pumps*. Paper C 49/72.

Osterwalder, J. (1978). Efficiency scale-up for hydraulic roughness with due consideration for surface roughness. *I.A.H.R.J. Hyd. Res.,* **16**, pp. 55-76.

Osterwalder, J. and Ettig, C. (1977). Determination of individual losses and scale effects by model tests with a radial pump. *I. Mech. E. Conf. Scaling for Performance Prediction in Rotodynamic Machines*. September, pp. 105-12.

Pearsall, I. S. (1966). The design and performance of supercavitating pumps. *N.E.L. Conf. Pump Design Testing and Operation.* Paper C 22.

Pearsall, I. S. (1970). Supercavitating pumps and inducers. Von Karman Institute Course on Pumps, December, pp. 14–18.

Pearsall, I. S. (1973). The design of pump impellers for optimum cavitation performance. *Proc. I. Mech. E.*, **187**, p. 667.

Pearsall, I. S. (1974). Cavitation. *The Chartered Mechanical Engineer*, July, pp. 79–85.

Pearsall, I. S. (1978). New developments in hydraulic design techniques. Pumps Conference: *Interflow 1978.*

Peck, J. F. (1968/9). Design of centrifugal pumps with computer aid. *Proc. I. Mech. E.*, **183**, p. 321.

Pfleiderer, C. (1961). *Die Kreiselpumpen.* Springer Verlag.

Rachmann, D. (1966). A Study of the Priming Process in a Centrifugal Pump M. Sc. Thesis, Glasgow University.

Rachmann, D. (1967). Physical characteristics of self priming phenomena centrifugal pumps. B.H.R.A. SP911.

Rayleigh, Lord. (1917). On the pressure developed in a liquid during the collapse of a spherical cavity. *Phil. Mag.* **34** August, pp. 94–8.

Richardson, J. (1982). Size, shape, and similarity. *I. Mech. E. Conf. Centrifugal Pumps - Hydraulic Design.* p. 29.

Ryall, M. L. and Duncan, A. B. (1980). The rectification of service failures in pumping equipment. *I. Mech. E. Conf. Fluid Machinery Failures.*

Salisbury, A. G. (1982). Current concepts in centrifugal pump hydraulic design. *I. Mech. E. Conf. Centrifugal pumps - Hydraulic Design*, 1.

Schwarz, K. K. (1985). Design of economical variable speed pumping systems. *9th B.P.M.A. Conf.* April, pp. 43–56.

Shilston, P. D. (1985). Designing pump drive systems for reliability. *9th B.P.M.A. Conf.* April, pp. 21–34.

Smith, A. G. and Fletcher, P. J. (1954). *Observations on the Surging of Various Low Speed Fans and Compressors.* N.G.T.E. Memeo. M. 219.

Stanitz, J. D. (1952). Some theoretical aerodynamic investigations of impellers in radial and mixed flow machines. *Trans. A.S.M.E.*, **74**, May, pp. 473–97.

Stepanoff, I. (1976). *Centrifugal and Axial Flow Pumps.* Wiley.

Stirling, T. E. (1982). Analysis of the design of two pumps using N.E.L. methods. *I. Mech. E. Conf. Centrifugal Pumps - Hydraulic Design.* pp. 55–73.

Stirling, T. E. and Wilson, G. (1983). A theoretically based C.A.D. method for mixed flow pumps. *8th Tech Conf. B.P.M.A.*, March, pp. 185–204.

Stodola, (1945). *Steam, and Gas Turbines.* 6th Edition. Peter Smith, New York.

Summers-Smith, J. D. (1988). *Mechanical Seal Practice for Improved Performance. I. Mech. E. Guide.* Mechanical Engineering Publications Ltd.

Sutton, M. (1967). *Inducer Design and Performance.* B.H.R.A. Report R889.

Sutton, M. (1968). Pump scale laws as affected by individual component losses.

I. Mech. E. Conf. Model Testing of Hydraulic Machinery and Associated Structures. April, paper 10.

Thorne, E. W. (1979). Design by the area ratio method. *6th B.P.M.A. Conference 'Pump 1979',* March, paper C2.

Turton, R. K. (1982). Fluctuating thrust loads in centrifugal pumps. *11th. Symposium of the Hydraulic Machinery Group of the International Association for Hydraulic Research.* Amsterdam, September.

Turton, R. K. (1984a). *Principles of Turbomachinery.* E. and F. N. Spon.

Turton, R. K. (1984b). The use of inducers as a way of achieving low NPSH values for a centrifugal pump. *World Pumps,* March, 77–82.

Turton, R. K. (1989). The gas liquid performance of a centrifugal pump: priming using the shroud reflux method. *11th B.P.M.A. Tech. Conf,* paper 9.

Turton, R. K. and Tugen, Z. (1984a). The inducer–pump combination, Part 1. *World Pumps,* May 157–163.

Turton, R. K. and Tugen, Z. (1984b). The inducer–pump combination, Part 2. *World Pumps,* September, 322–32.

Varley, F. A. (1961). Effects of impeller design and surface roughness on the performance of centrifugal pumps. *Proc. I. Mech. E.* **175,** 955.

Waldron, W. J. and Jackson, D. (1983). A system for achieving variable flow from fixed speed pumps. *8th B.P.M.A. Conference,* March, 67–86.

Ward, T. and Sutton, M. (1958). *A Review of the Literature on Cavitation in Centrifugal Pumps.* B.H.R.A. TN 892.

Watabe, K. (1958). On fluid friction of rotating rough disc in rough vessel. *J.S.M.E.,* **1,** (1), pp. 69–74.

Weinig, F. (1935). Die Stromung um die Schaufeln von Turbomaschinen. *Joh. Ambr. Barth.* Leipzig.

Weisner, F. J. (1967). A review of slip factors for centrifugal impellers. *A.S.M.E. J. Eng. for Power.* Oct: 894 (4) p. 558.

Wilson, G. (1963). *The Development of a Mixed Flow Pump with a Stable Characteristic.* N.E.L. Report 110.

Wilson, J. H. and Goulburn, J. R. (1976). An experimental examination of impeller passage friction and disc friction losses associated with high speed machinery. *Int. Conf. on the Design and Operation of Pumps and Turbines.* N.E.L., Sept. paper 5.1.

Wisclicenus, G. (1965). *Fluid Mechanics of Turbomachinery,* (2 volumes). Dover Press.

Wood, M. D. (1986). Streamline curvature methods. Cranfield Fluid Engineering Short Course Advanced Flow Calculations for Internal Flow and Turbomachinery.

Wood, M. D. and Marlow, A. V. (1966). The use of numerical methods for the investigation of flow in water pump impellers. *Proc. I. Mech. E.,* **181,** part 1, no 29.

Worster, R. C. (1956). Cavitation. *BHRA 4th Conference on Hydrodynamics,* Leamington.

Worster, R. C. (1963). The flow in volutes and effect on centrifugal pump performance. *Proc. I. Mech. E.*, **177**, 843.

Wu, C. H. (1952). *A General Theory of Three-Dimensional Flow in Subsonic and Supersonic Turbomachines of Axial, Radial, and Mixed Flow Types.* N.A.C.A. T.N. 2604.

Young, F. R. (1989). *Cavitation.* McGraw-Hill.

Index

aerofoil
  definitions, 62–3
  properties, 65–9
  theory, 63–5
affinity laws (scaling laws), 7–8
air ejector, 174
angle
  inlet, 3
  of incidence, 63
  outlet, 3
API 610, 182–5
area ratio, 50–1, 104
axial flow impeller, 6, 13, 128
  pump, 4, 60, 124
axial thrust
  in axial pump, 79
  in centrifugal pump, 55–9
axial thrust balance, 58–9

balance chamber, 55
balance disc, 59
balance drum, 59
bearings
  hydrodynamic, 145–7
  rolling element, 142–5
blade loading, 137
blade proximity correction, 135
boiler feed pump, 32

camber, 62
canned pump, 169
cascade data, 69–73
casing
  axial, 137
  bulb, 139

double volute, 55
  inlet designs, 31–5
  volute, 117–19
cavitation
  criteria, 19–20, 23
  damage, 17
  design criteria, 24–6
  effects, 18–19
  inception, 15
  noise, 23–4
  resistance, 18
  scaling, 26
C4 profile, 67
centrifugal impeller blade design, 108–14
characteristic number (or specific
    speed), 10
chemical pumping, 165–71
Clark 'Y' profile, 69
classification of pumps, 10–13
clearance (wear ring), 184
coefficient
  drag, 64
  flow, 6
  lift, 64
  head, 6
  power, 6
computer based techniques, 82–94
conversion factors (specific speed), 12
corrosion resistance, 163
critical speed, 182

design process, 181
deviation angle, 134
diameter ratio, 11
dimensional standards, 181–5

195

dimensionless groups, 6-7
disc friction, 14
discharge casing, 47-53
double suction pump, 31, 35
drag coefficient, 64
drag force, 63

efficiency
  hydraulic, 10
  mechanical, 10
  volumetric, 12
end suction pumps, 30
  design, 95-123
erosion resistance, 163
Euler equation, 2
  axial flow pump, 5
  centrifugal pump, 3-4

flat plate data, 70
flow
  coefficient, 6
  deviation, 133
  free vortex, 75
  recirculation, 33, 36
free vortex law, 75

gas-liquid pumping, 171-8
glandless pumps, 168-71
Gottingen profiles, 66

head coefficient, 6
  Euler, 2
  shutvalve, 97
hot/cold duties, 166
hub section design (axial), 132
hub-tip ratio, 130

impeller
  axial, 128-37
  centrifugal, 108-19
  mixed flow, 138-9
inducer, 27-8
inlet casing, 31, 34-5
isolated blade data, 62, 65-70

leakage flow, 13
leakage loss, 13
lift to drag ratio, 132
liquid ring pump, 174
losses
  disc friction, 14
  hydraulic, 12
  leakage, 13
  mechanical, 12
  recirculation, 35
  volumetric, 13

material damage, 17, 18, 163
mechanical losses, 12
mechanical seals, 147-52
  single, 147-50
  double, 150-2
  selection, 152
mixed flow impeller, 77-8
multi-stage pumps, 29, 32, 52-3
  design, 122

NACA profiles, 70-3
NEL axial/mixed flow design package, 78, 139
net positive suction head (or NPSE)
  available, 20-1
net positive suction head (or NPSE)
  required, 20-1
NPSE critical, 21-2

optimum suction (or eye) diameter, 24-6

packed gland, 153-7
  materials, 153
  housing, 154-6
  performance, 157
peripheral velocity, 3
pre-rotation, 34-5, 36
pressure coefficient, 6
priming systems, 171-7
process pumps, 161
pump
  bearings, 142-7
  casing, 47-53, 117-19, 138
  characteristics, 5, 7-9, 27
  design for low NPSE, 24-6
  drives, 157-60
  losses, 10-14
  neck (or wear) rings, 120-2
  priming, 171-7
  seals, mechanical, 147-52
  seals, packed gland, 153-7
  testing, 185

radial equilibrium, 73-5
recirculation, 35-6
Reynolds number, 6
rotating stall, 76
rotational speed (choice)
  axial, 128
  centrifugal, 30-1, 95

scaling laws, 7-8
seals
  single mechanical, 147-50

double mechanical, 150-2
 packed gland, 153-7
shaft design, 140-1
shroud thickness, 116-17
shut-off head, 4, 97-8
slip, 44-7, 107-8
slip factors, 45, 46
solidity, 132
solids handling pumps, 168
specific speed, 9-10
stagger angle, 63, 136
stall, 75-6
straight conical inlet, 34
stream surface, 77, 81
stuffing box, 153-7
suction geometry, 24-6, 99
suction specific speed, 23

Thoma's cavitation index, 23
three-dimensional flow effects (axial),
 73
throat area (volute), 117
thrust forces

axial, 55-9, 79, 119-20
 radial, 53-5, 119

vaned diffuser, 52
vapour pressure, 20
velocity
 absolute, 2
 distribution, 44, 73
 ideal triangles, 2-6
 peripheral, 2
 whirl, 2
vibration, 141, 182, 184
volute
 casing, 47-53
 casing design, 117
 double volute design, 55
 throat area, 117

wear ring (or neck ring)
 clearance, 121, 184
 designs, 120-2
Weisner's slip equation, 45-6
whirl velocity, 3